RAW ENERGY

Award-winning writer and former television broadcaster Leslie Kenton was Health and Beauty Editor of *Harpers & Queen* for fourteen years. A shining example of health and vitality, she has been described as 'the most original guru of health and fitness'.

Leslie lives part of the time in London and part of the time in Pembrokeshire, Wales—in a 250-year-old house overlooking the sea.

Susannah Kenton is her daughter, who after obtaining a degree from Columbia University in the USA went on to study drama in London. She now lives and works in Paris.

Both mother and daughter are supreme examples of what they teach. Both follow a 75 per cent raw-food diet which has transformed their lives.

RAW ENERGY

Leslie and Susannah Kenton

ARROW

Arrow Books Limited
20 Vauxhall Bridge Road, London SW1V 2SA

An imprint of Random House UK Ltd

London Melbourne Sydney Auckland
Johannesburg and agencies throughout
the world

First published by Century 1984
Century paperback edition 1985
Century Arrow edition 1986
Arrow edition 1987

15 17 18 16

Printed and bound in Great Britain by
Cox & Wyman Ltd, Reading, Berkshire

ISBN 0 09 946810 7

for
CHRISTI and CALVIN

ACKNOWLEDGEMENTS

We would like to express our gratitude and deepest appreciation to the following people who have been so helpful during the research and preparation of this book: Dr Ralph Bircher, Dr Dagmar Liechti von Brasch, Dr Gordon Latto, Dr Barbara Latto, Dr Phillip Kilsby, Dr John Douglass, Dr Alan Clemetson, Dr Henning Karstrom, Dr Chiu-Nan Lai; the directors of the Price Pottenger Foundation, the Lee Foundation for Nutritional Research and the Cancer Control Society; Arden Perrin, Angelika Langoosch, Jacquie Atkinson, Nicola Heywood, Monica Fine and Graham Nicholson.

CONTENTS

FOUR
RAW ENERGY RECIPES

FOREWORD

'Raw energy' is a zesty title for this exciting and important book. Exciting because the ideas are as freshly presented by this mother and daughter team as the food they recommend, and important because they describe a whole science of healthy eating.

Healthy eating is a timeless science, researched for thousands of years in different civilisations but surprisingly neglected this century. Here we are given the full history of the use of a raw food diet in what ancient Taoist tradition considered the goal of life – achieving harmony. Streams of thought in religious, philosophical and medical teaching have given rise to many different ideas for balancing the opposing aspects of human life. To some they represent the dynamic, often destructive, male principle seen against the constant nurturing female principle, the shiva versus shakti of Indian philosophies, yang versus yin of Chinese traditions. Modern scientists would see them as catabolic and anabolic, acid or alkaline producing metabolic processes, and sympathetic fight and flight or para-sympathetic rest and digest divisions of the autonomic or automatic nervous system preparing the body for war or peace. As the theoretical physicist Fritjof Capra describes in his recent book *The Turning Point*, we are now beginning to appreciate the urgency of the need to move towards a more balanced mid-point between these polarities if we are to avoid disaster at both the individual and global level.

This book shows how food could help the individual person to achieve this balance, producing less acidity and more alkalinity in the blood, and even calming the mind to lessen the impact of stress. As this book graphically describes, malnutrition is not just insufficient food or the wrong food, but it can be that the natural energy present in its fresh raw state is destroyed by cooking or preserving.

Truly it has been said that we shall not live by bread alone, even when vitamin and mineral enriched!

The theme of a raw food diet is very topical in these days when the public are demanding a 'holistic' approach to medicine. Nutritionally aware physicians are few and far between, especially when it comes to factors such as the subtle energies of living plants and plant enzymes which are so well covered by the Kentons.

Doctors are waking up to the benefits of a high-fibre diet in preventing a range of disorders from heart disease to bowel cancer and, largely through the work of the McCarrison Society, even psychiatrists are becoming aware of the 'Psychopharmacology of Food', as one recent conference at the Royal Society of Medicine was called. However, like their patients and the public at large, they could all greatly benefit from the food for thought which is so tastefully served at a practical as well as at a theoretical level in this book.

Malcolm Carruthers
November 1983

Dr Malcolm Carruthers MRCPath., MD, MRCGP is the Director of Clinical Laboratory Services and Consultant Chemical Pathologist at the Maudsley and Royal Bethlem Hospitals. He is also Director of the Positive Health Centre in London.

INTRODUCTION

When people asked us about the book we were writing together, we told them it was about raw food. Many looked at us in disbelief. We were, they implied, either slightly mad or terribly cranky. Some questioned why, 'when there are so many valuable subjects to write about' were we 'wasting our time investigating the effect on health of munching carrots'. Others smiled indulgently and maintained a polite silence.

We could see their point. Our fascination with uncooked foods could seem very eccentric to anyone who knew nothing about the enormous quantity of European research into the health-promoting effects that uncooked foods have on the body or someone who had never experienced the energising effects of a raw diet. Our delving through literally hundreds of books and research papers on the biochemistry and the clinical uses of uncooked foods would appear rather strange to someone totally unaware of the vast quantity of evidence which exists showing that the high-raw diet – a way of eating in which 75 per cent of your foods are taken raw – can not only reverse the bodily degeneration which accompanies longterm illness, but retard the rate at which you age, bring you seemingly boundless energy and even make you feel better emotionally. We asked ourselves could such a way of eating also offer a freedom from the misery and serious illness which now afflict a majority of people aged over 40 (and many much younger) in the industrialised world? We knew that a high-raw diet of foods grown on healthy soils and eaten fresh is without question the finest complement of essential nutrients – both known and unknown – that you can find in nature. On this even the most conservative nutritionists agree. Could the excellent natural synergy of nutrients in raw foods, we wondered, be the answer to their potent ability to improve health? That is where our search for information began.

Our aim was simple: to gather together all the facts we

could find about raw food effects. Our research project became very exciting. For, just as we would think we had an understanding of the central issues involved in the therapeutic actions of uncooked foods, we were surprised by the discovery of a whole new collection of information that made the picture even more complex. Before we closed the final page of our investigations we discovered that there is far more to the health-giving properties of uncooked foods than merely their high vitamin and mineral content. Some of the facts we uncovered along the way about how they affect the body have actually begun to alter our whole way of looking at health and life. More about that later.

This book is not a work of disinterested scholarship. Its impetus and meaning for both of us came from two things. First, in experimenting with various ways of eating over many years, we, like many others, have found that a diet that is about 75 per cent raw makes us look and feel best, gives us high levels of energy and stamina to cope with day-to-day living and protects us from minor illnesses such as 'flu and colds. (Neither of us has ever had a major illness.) Second, so profound were the energising effects that a high-raw diet had on us that (having grown up in the twentieth century with its passion for scientific explanations of phenomena) we were fascinated to know why. What was so special about uncooked foods? What gave them the power to heal even long-standing illnesses? What was it about a diet of raw foods and juices – the typical regime at the world's most exclusive health spas – that had men or women looking ten years younger after a fortnight on such a diet?

At first we thought that the effects of the high-raw diet were entirely personal – some kind of genetic penchant we had both inherited from a long line of rabbit-like ancestors which made us biologically susceptible to what we had come to call 'the raw food effect'. Then we looked around us. We spoke to numerous physicians on both sides of the Atlantic who had long experience in the treatment of illness with uncooked foods. We began to delve through libraries and to seek out old volumes – and volumes not-so-old – which recorded similar experiences to ours – some far more dramatic. Much of what we read about healing with these simple substances far surpassed anything we ourselves had experienced, since neither of us had been really ill. We discovered that the Germans, and the Swedes and the Swiss have for generations catalogued the healing effects of a diet high in fresh raw vegetables and fruits. We found that such a diet had been credited with the healing of long-term crippling diseases such as arthritis and cancer, gastric ulcers, diabetes and heart disease. We uncovered reports that athletes taken off their usual diet high in cooked animal protein and put onto a raw regime not only lost none of their physical prowess but could actually improve their performances.

We began to ask more questions. We wanted to know why these uncooked foods were capable of working such 'miracles'. It was not an easy question to answer, for clinical reports and the biochemical research into the actions of raw foods on an organism span almost a century. They are complex, sometimes contradictory and usually in foreign languages.

Finally we wanted to discover, and then set out in some practical and usable form, guidelines for anyone who, like ourselves, was interested in discovering how a high-raw diet can help explore the 'further reaches of human health'. By this we mean a way of living where you awaken in the mornings feeling fresh and good about yourself and your life – a state of being where your physical and mental potential has the best chance of being used in your work – that realm of consciousness in which your capacity for fun approaches that of a joyous child who finds delight in everyday experiences and a sense of excitement in each challenge. For we believe that real health is not just the absence of disease. It is a dynamic state of mind and body that makes it possible for you to participate spontaneously in the fullness of life.

Those who are optimally healthy do not face the future with fears about getting old and falling prey to one of the many degenerative diseases. Rather, they have a kind of gentle anticipation and excitement because they know they are capable of moving toward an even more positive state of health and fulfilment as the years pass. They have a sense that 'the best is yet to come'. They are no longer ruled by the old model of aging that says getting older includes illness as a natural part of life. They know different. Their paradigms of aging are closer to that of world famous age researcher Johan Bjorksten who speaks of giving 'as many people as possible as many more healthy vigorous years of life as possible'.

We are not scientists – we are merely reporters. We make no claims for healing with raw foods. Our aim is to report on the research and opinions of scientists – many of them Nobel Prize Winners – working in studies related to the actions of uncooked food on health and to share with you some of our own experiences with a high-raw diet. We hope that the information we have gathered together will be of interest. This information, however, is in no way intended to be prescriptive. No book can replace adequate medical care. We would suggest that any reader who is unwell seek the attention and care of a nutritionally aware physician.

The book is in four sections. The first deals with the scientific research and clinical experience of scientists using raw foods for health. You can either read it all through for an overall view or simply thumb through chapters dipping into areas of particular interest. We have given a full list of references relating to each chapter on pages 289–301 for those who would like to explore further for themselves. The second section tells you some of our own experiences with a high-raw diet and outlines how a high-raw diet can help retard the aging process, alleviate stress and cope with women's problems or how it can be useful in building high levels of mental and physical energy. In the last two sections we offer a practical guide for those people simply looking for the answer to the questions: 'How do I begin exploring the potentials of a high-raw diet for myself?' and 'What exactly do you eat when you want to eat mainly uncooked foods?'

We hope that you will use the book to suit your personal needs and interests.

We believe that all people have far more resources for happiness, health and creativity than they generally use. The high-raw diet has brought us to a high-energy state of health which delights us and which has us looking towards the future with enthusiasm. We'd like to share some of our research, our experience, and our enthusiasm with you.

ONE

COOKING
MAY DAMAGE
YOUR HEALTH

1 MESO-HEALTH OR SUPER-HEALTH?

Experimenting with animals at the University of Rostock just before World War II, Professor Werner Kollath discovered an extraordinary phenomenon: he called it 'meso-health', which means 'middle-nourishment'. He found that a diet which did not produce high-level health nonetheless sustained 'normal' health. He took animals and reared them on a diet of purified, processed foodstuffs devoid of all minerals except potassium phosphate and zinc and virtually empty of vitamins except a little thiamin. Despite their poorly nourished condition his animals grew and showed no clinical signs of disease, not even vitamin deficiencies.

Later on, however, by the time they reached adulthood, they showed signs of degeneration similar to those common among humans in the Western industrialised world – dental caries, constipation, large populations of harmful bacteria in the colon, and loss of calcium from their bones. Autopsies of their vital organs revealed damage similar to that found in humans suffering from degenerative diseases. No amount of vitamin supplements appeared able to reverse these unpleasant changes. The only thing that did, provided it was given early enough, was an abundance of fresh raw food containing green leaves, cereals and vegetables. Kollath's findings were later confirmed by researchers working in Stockholm and Munich.

Scientists involved in raw food research and in the treatment of illness using raw food diets believe that a great many people in the industrialised world are, like Kollath's animals, probably living in a state of 'meso-health', a state of half-hearted health induced by years of eating devitalised and processed foods. They say that by increasing the quantity of fresh raw food we eat, and cutting out many of the cooked, processed, high-protein and high-fat foods, it would be possible to improve the health of people already suffering from degenerative illness and also the health of those in whom the subtle processes of degeneration are not yet obvious. A lot more research needs to be done to establish the wide-scale health-promoting effects that such an approach would have. One thing is certain, however: one cannot exaggerate the current state of poor health in Britain and America.

People are not getting healthier

Despite a sophisticated medical system based on potent drugs, dramatic lifesaving techniques and high technology procedures such as coronary bypass surgery, kidney dialysis and prosthetic joints, the state of health of most people in Britain and the United States is poor. The diseases from which we suffer – cancer, cardiovascular troubles, diabetes, arthritis, respiratory disorders such as emphysema and bronchitis, and depression – have shown little decline in their incidence since the turn of the century. Indeed most of them, like cancer and mental illness, are steadily increasing.

In Britain some 200000 cases of cancer occur each year. In 1981 the American National Cancer Institute predicted that one in every three Americans will get cancer before the age of 74. The situation with heart disease is not much better. Dr Robert Levy, director of America's National Heart, Lung and Blood Institute, has reported that some 35 million Americans suffer from high blood pressure, the primary cause of one and a quarter million heart attacks and half a million strokes every year. Here in Britain more than 150000 people die every year from strokes and heart attacks. Ill health in Britain costs the taxpayer about £12 000 million a year – the annual budget of the National Health Service – three times as much in real terms as in 1949. In the United States medical costs have risen from $27 billion to $200 billion a year in the last twenty years. In short, we are not getting any healthier, and the health we have is costing us a fortune. That is quite apart from the suffering involved, which cannot be computed in money terms.

Now that so many viruses and bacteria have been tamed

thanks to widespread vaccination, improved public hygiene and the discovery of antibiotics, the diseases to which late twentieth century Western society is heir are mainly degenerative: heart disease, hypertension, circulatory ailments, cancer, diabetes, arthritis, obesity, hypoglycaemia, mental disorders. These are illnesses we have acquired along with our twentieth century lifestyle. They are the result of stress, overeating, under-exercising and pollution of our environment, whether it is radiation, airborne chemicals or chemicals in our food that are doing the polluting. High technology medicine and eleventh hour intervention can do little to prevent or cure such illness. As a first positive step towards better health we must stop treating our bodies with what Kenneth Pelletier, author of *Holistic Medicine*, has dubbed the 'Volkswagen attitude'. We must give up the notion that we can run ourselves as long and as hard as we like and then expect the doctor to pick up the pieces or supply us with new parts when we break down.

The new ecological approach

Ecology is now the watchword of all the life sciences, and it is time we started applying it to health. We must acknowledge that the body, like the earth itself, has finite resources, that the condition we find ourselves in after two or three or four or five decades of living depends primarily on how we have lived. More specifically, it depends on a number of variables over which, ultimately, only *we* can exercise control. These variables include such things as how we cope with stress, how often and to what degree we subject our bodies to drugs, alcohol and cigarettes, how much physical exercise we take, and what levels of environmental pollution we allow our governments to pass as 'safe'. But probably the single most important variable of all is nutrition.

Eating habits are relatively easy to change, much easier to change than legislation about lead emissions in car exhausts, for example. And research indicates that changes in how and what we eat very significantly affect health. Numerous recent studies show that when nutrition improves so does health. Even small changes, such as cutting down on processed foods, sugar, alcohol and fat, can improve health relatively quickly. Changes for the better are measurable. They show up as lower levels of cholesterol and triglycerides in the blood, lower blood pressure, and more efficient functioning of the immune system which protects the body from infection, malignant growths and early aging. More important, even minor changes in diet profoundly affect how we look and feel, and to most people that is the most important 'proof' that what they are doing is right for them.

Outmoded nutritional thinking

The notion most doctors have of nutrition is naive and incomplete. It is largely based on nineteenth century scientific thinking which assumed that disease A was caused by nutritional deficit B – the single cause/single effect model of disease. Even today standard academic texts on nutritional diseases confine themselves to the 'deficiency diseases' – scurvy, beriberi, pellagra, rickets and kwashiorkor. The list is short and the diseases are those that are caused by a shortage of one specific nutrient in the diet. In reality these gross manifestations of deficiency are rare in developed countries today. Sub-clinical deficiencies which cause slow but inexorable physical decline are not.

It is also common for the average doctor to insist that provided a diet is 'well-balanced' – and for most people the balance is between one highly processed food and another – it offers enough nutritional support for health. This is simply untrue, as has been shown in a number of large scale, well conducted studies both in England and the United States. The three-year Health and Nutrition Examination Survey carried out between 1971 and 1974 on 28 000 people in America by congressional mandate is an excellent example. Even by highly conservative estimates it showed that half the women surveyed had calcium deficiencies; that iron deficiency was widespread among people of every race, income group and cultural background; and that more than

60 per cent of those examined had at least one symptom of malnutrition.

Nutrition and evolution

Between 50 000 and 100 000 different chemicals go into the making and running of the human body. They interact with each other in ways so complex that they make the world's most advanced computer look like an abacus. Nutritional science has so far isolated and identified some 17 vitamins and co-factors, 24 minerals, and 8 to 10 amino acids as being essential to the health and reproductive abilities of the human body. No doubt there are many more – new substances are discovered relatively often. Health and fitness depend greatly on the quality and variety of the nutrients we take in. Without them we could not build and maintain the elaborate machinery and perform the complex chemical transformations on which life depends.

These essential nutrients, far from being isolated in their workings, are synergistic, which is a way of saying that they need each other. It is only by working together that they can enact the complicated routines that make our bodies function. The human body has evolved to make use of the myriad co-operative and complementary substances which occur in foods in their natural state. There are many similarities between human blood and substances and fluids found in nature; for instance, blood serum has a composition not unlike that of sea water, which is a veritable consommé of minerals; the blood pigment haemoglobin has a molecular structure that closely resembles the plant pigment chlorophyll.

Foods in nature are highly complex; as the great Soviet biochemist Professor I. I. Brekhman says, they are 'rich in structural information for health'. Cooking and other forms of processing interfere with this complexity and destroy much of this structural information. To quote Michael Colgan, author of *Your Personal Vitamin Profile*: 'The multiple interaction of these essential substances is the basis of their biological function. And the adequacy of that function depends on the substances being supplied to the body in the same mixtures and concentrations that occur in raw, unprocessed foods. It was by use of these foods that genesis, over millions of years of evolution, developed the precise mechanism to deal with them.'

It is not surprising that a diet which deviates too far from what our bodies have been genetically programmed to expect leads to progressive malnutrition. First there is deprivation at the cellular level, then gradual failure of the body's immune system, then illness. Another of the West's most highly respected authorities on nutrition, the American biochemist Roger Williams, has said that cellular malnutrition is at the root of ten times the number of disease conditions as clinically defined deficiencies. The diseases Williams is referring to include allergies, arthritis, atherosclerosis (hardening of the arteries), coronary heart disease, emotional disorders, insomnia, peridontal (tooth-related) diseases, infections, skeletal deformities, mental retardation and disorders of the immune system. If we are to overcome the current crisis in Western medicine, he says, we must find ways of tackling malnutrition at the cellular level. That means altering the focus of our effort away from treating the symptoms of illness and looking for outside causes. Instead we must concentrate on overall health and vitality and strengthening the immune system. The easiest, most effective way of doing this in the long term is by improving nutritional habits.

The all-important immune system
The immune system is the number one suspect in degenerative disease and in premature aging. It is really split into two interdependent parts – the thymus with its T-lymphocytes or T-cells, which is the main system of cellular immunity, and the thymic-independent B-lymphocytes or B-cells which protect us from most viral and bacterial infections.

Numerous studies have shown that the nutrients which occur in optimal proportions and quantities in fresh uncooked vegetable foods – particularly in home-grown sprouts and vegetable juices – boost lymphocyte production and so increase resistance to illness. The nutrients that most favourably influence the immune system are vitamins E, C and A, many of the B vitamins, and also zinc. Without a full complement of these fresh food nutrients (and probably others not yet tested and some perhaps not yet even discovered) immunity cannot be maintained, and we condemn ourselves to a state of mesotrophy. The body has quite extraordinary powers of compensation of course; for many years one can eat the wrong foods and show no clinical sign of illness. But the insidious degenerative process is at work within. Sooner or later, depending on one's constitution and how badly the body is abused, hidden degeneration turns into serious illness.

The self-regulating system

From our research into the clinical uses of diets high in raw foods for healing and into the biochemical studies that suggest why raw food is so potent a force for health – and also from our personal experience – we have come to believe that a high-raw diet offers two precious things: a promise of greater resistance to illness and aging, and a key to greater 'aliveness' and vitality. High-level health can only come from a finely tuned, self-regulating, self-enjoying system. If you are aware enough of your own needs and responses you become the 'authority' on what is best for you.

No one living in what the great Swiss physician Max Bircher-Benner called 'the twilight zone of ill health' – mesotrophy by another name – has such a possibility. When an organism's vitality is lowered, or when its biochemical balance is disturbed, messages from the systems of the body to the conscious mind become garbled. How can self-monitoring and self-correcting mechanisms be expected to work if both physical and mental perceptions are distorted? Perhaps the greatest satisfaction a raw diet has to offer is the experience of finely tuned self-regulation day after day, year after year. 'Spiritual' is not a fashionable word today. Nevertheless we believe that the benefits of a high-raw diet are of a spiritual as well as a physical order.

The American physician Dr John Douglass discovered a strange pattern when talking to patients at the Kaiser-Permanente Medical Center in Los Angeles; a significant number of them said that after a few weeks on a high-raw diet they felt 'declimatised to Western life'; they found old habits such as cigarettes and alcohol distasteful. This suggests that a high-raw diet tends to make the body more sensitive to whatever it is exposed to. Sexually and aesthetically that could be very satisfying. Higher sensitivity to all kinds of stimuli would make it easier to judge 'instinctively' what does you good and what does you harm. If you receive and heed those signals you intuitively do what is best for you. But this is just the beginning of the amazing Raw Energy story.

2   RAW FOOD PIONEERS

Ironically many of the most enthusiastic pioneers of raw foods discovered their health-promoting qualities as a result of personal health problems – illness which only raw eating appeared to cure. In most cases they attracted the opprobrium of their professional colleagues and the gratitude of many hundreds of patients who had expected to be sick for the rest of their lives.

The Swiss physician Max Bircher-Benner, born in 1867, was one of the great European pioneers of nutritional science. He came upon the potential of uncooked foods when, as an overworked young doctor, he was seized with an attack of jaundice. It sent him to bed for several days and made it impossible for him to eat anything. His wife, peeling apples for dinner one day, slipped a small piece of the raw fruit between his lips. He found it pleasant and, unlike all the other food that had been offered to him, digestible.

Several days and many well chewed apples later he had completely recovered.

Soon afterwards he was called on to treat a patient who was apparently unable to digest anything. She was slowly starving and very weak. He mentioned the case to a colleague who happened to have an interest in ancient history. Did he know, his colleague mused, that in 500 BC Pythagoras wrote of curing a similar condition by giving nothing but mashed raw fruit, a little honey and goat's milk? Bircher-Benner was sceptical, despite his own experience with raw apples. Such a 'cure' broke all the rules. He had been taught, as are most health professionals today, that raw foods are difficult to cope with if you have an ailing digestive system. But as everything else had failed he decided to try Pythagoras' remedy. His patient ate the raw foods he gave her and tests the next day showed that she had digested them properly. A digestive system that could not handle cooked food thrived on raw food. Spurred on by his patient's remarkable return to health Bircher-Benner began to investigate the special properties of 'living foods' as he called them and to use them to treat other illnesses. Regardless of the type or seriousness of the illness his living food treatments were an enormous success. The clinic which he founded in Zurich in 1897 continues to be one of the most highly respected centres for healing in the world.

Bircher-Benner foresaw all too clearly the epidemic proportions that the degenerative diseases have now reached in the industrialised West. Not long before his death in 1939 he wrote:

'. . . we are oppressed by an overwhelming burden of incurable disease which hangs over our lives like a dark cloud. It is a burden which will not disappear until men become aware of the basic laws of life. As it is we doctors have to concentrate so much of our attention upon the task of keeping the incurable alive with the aid of artificial "crutches" that the divinely ordained role of our profession – the healing of the sick and the prevention of disease – is forced more into the background. Neither the profession nor the public seems to see the tragedy of this situation.'

Bircher-Benner was a proponent of 'holistic' medicine long before the term was coined. He insisted that a patient be treated as an indivisible whole – a psychophysical personality – with the end in mind not only of curing illness but also of promoting the realisation of the person's full potential. He saw each human being as unique, as having been born with a 'blueprint' that had to be realised. And he found that a diet high in uncooked foods along with regular exercise played a central role in this self-realising, self-healing process.

The universal cure?

The ailment that sent the German physician Max Gerson, a near contemporary of Bircher-Benner, in search of food remedies was migraine. Severe migraine ran in his family and at times his headaches, and the nausea that accompanied them, sentenced him to lying in darkened rooms doing nothing for days on end. The medical experts of the day told him there was no cure for migraine. Probably his headaches would disappear by the time he was 40 or 50.

Gerson was too young and impatient to wait that long, and so started to experiment with his diet. First he tried milk, reasoning that since it is perfect for babies it might be healing for him and easy to digest. But on a milk diet he felt sicker than ever. Then he turned to fruit. If his monkey ancestors had lived healthily on fruit and nuts and green vegetables, so could he. He began with apples, and cautiously extended this basic regime to include other fruits. His headaches stayed away, except when he added new items that disagreed with him. He ate fresh fruit and vegetables for the rest of his life. Gerson rose to be, in the opinion of his one-time patient Albert Schweitzer, 'one of the most eminent geniuses in medical history'.

When Gerson apologetically suggested his apple diet to a young man who came to him suffering from migraine he did not expect dramatic results. Nevertheless the young man reported that on the diet his headaches vanished too. He reported something else as well: another ailment from which he suffered, a kind of tuberculosis of the skin called lupus, had also disappeared. Gerson insisted that this was impossible. Lupus was completely incurable. Nobody in the history of medicine had ever reported a lupus lesion healing. Yet on the migraine diet other lupus sufferers sent to Gerson by the same patient also got well. Gerson treated them all without charge, hardly daring to claim that his diet was a serious therapy. Then in 1928 Albert Schweitzer's wife came to Gerson suffering from severe lung tuberculosis. She too made a complete recovery on his migraine diet. It was then that Gerson truly began to believe that his diet was much more than a cure for migraine. It was a way of eating that restored an ailing body's ability to heal itself. That explained why it was so effective across a whole range of ailments.

Although Gerson went on to use uncooked fresh foods and juices pressed from raw vegetables and fruit to treat everything from mental disorders to coronary heart disease, he finally became most famous for his treatment of cancer. His book *A Cancer Therapy: Results of Fifty Cases*, first published in 1958, is still the vade mecum of all physicians using natural and metabolic treatments for cancer. Gerson believed that the starting point for all illness, cancer included, is an imbalance of sodium and potassium, usually too much sodium and too little potassium. If you correct that balance by eating potassium-rich raw foods, which invigorate and cleanse the body because they improve respiration at the cellular level, you also mobilise the white blood cells which fight and destroy cancer cells. Gerson's raw food regimes enable many 'terminal' cancer patients to fight and destroy their cancers.

Physician, heal thyself

Many other physicians and healers have cured themselves of serious illness by eating uncooked foods. As a last resort Danish physician Kristine Nolfi turned to raw foods in an effort to beat breast cancer. She won. Then she taught her patients about this natural form of healing. Her success was so great (as was the fury her treatments unleashed among her orthodox peers) that she gave up using drugs altogether and started the Humlegaarden sanatorium in Denmark which she directed until her death in 1957. Since then her work has been continued by Dr F. Skott Andersen.

The American expert on raw juice therapy Dr Norman W. Walker, now 107, rid himself of the excruciating pains of neuritis with raw foods and then went on to write numerous books on how to use them for high-level health. Naturopath Ann Wigmore, founder of the Hippocrates Health Institute in Boston, is now in her seventies and travels the world lecturing and writing about the raw foods which transformed her life when she was in her fifties and seriously ill. The German scientist Arnold Ehret suffered from heart disease, kidney trouble and Bright's disease until he discovered that fasting and fresh fruit cured him of all those things, and at a fraction of the cost of all the treatments he had tried previously. Convinced of the health-destroying effects of the wrong kind of food, in particular of the 'internal pollution' they cause in the colon, he developed the famous Mucusless Diet which he taught to wide acclaim in Europe and America until his accidental death in 1922.

'Primitive' diets

Another pioneering advocate of a high-raw diet was the American dentist Weston A. Price. From 1920 to 1940 Price travelled the world studying primitive societies – looking at the development of teeth and bones, the incidence of dental caries, and the general physical and mental health of isolated cultures. He examined the dietary patterns of many different peoples, from the Loetschental Valley people high up in the Swiss Alps to the rugged Gaels of Harris in the Outer Hebrides. The outcome was a fascinating book, *Nutrition and Physical Degeneration*, published in 1945, in which he carefully documented his findings, complete with photographs and statistics. His main conclusion was bleak indeed: processed foods pose appalling dangers to human health. He argued that human health is related to the wholeness and freshness of the foods we consume, and that a high level of health is almost impossible to achieve unless a diet is rich in uncooked foods.

Price discovered that, despite great differences in the specific foods they ate, the diet of peoples who were largely free of mental and physical diseases, had good skeletal structure and few, if any, dental caries, had many things in common. It was what he called a 'primitive diet', a diet of simple, fresh and largely uncooked foods which were usually gathered and used immediately. These peoples spread natural fertilisers on their crops and knew nothing of fungicides or insecticides.

By contrast our 'modern diet' consists of an enormous variety of foods, many of which are tinned, frozen or preserved, and even our fresh foods are adulterated – lettuce is plunged into chemicals such as N-6 Benzyladenine to keep it fresh and many of the apples we buy have been sprayed with as many as 18 chemicals. The staples of the modern diet are different too. They are not the fresh fruit and vegetables, seeds and grains of Price's isolated cultures. They are the protein foods – meat, fish, poultry and dairy products made from pasteurised milk – and massive quantities of refined flour and sugar. And we take in several pounds of chemical food additives each year, the effects of which have been strongly questioned even by those professionals who still preach the standard 'well balanced' British and American fare of meat-and-two-veg.

When Price visited the remote Loetschental Valley in Switzerland in 1932 its 2000 inhabitants had only one route to the outside world, a winding single-track railway. Examining their records, which went back some two hundred years, Price found there had never been a case of tuberculosis among them. They had no policeman or gaol, or doctor or dentist – there had never been a need for them. Price saw very clearly that diet is more than physiological in its effects; it strongly influences behaviour and environment too.

His fieldwork done, Price went back to America advocating a return to simple fresh foods grown in organically fertilised soils. He found, like Gerson and Bircher-Benner and so many other advocates of dietary change, that his colleagues ostracised him. His approach was too radically simple and too ecological to be swallowed by a scientific community committed to high technology procedures and intent on reducing disease to a matter of single causes and single effects. Yet, like the findings of McCarrison over the past 40 years, Price's intuitions have been increasingly confirmed by epidemiological research all over the world.

Pottenger's amazing cats

Further confirmation of Price's findings came from American physician Francis M. Pottenger. His laboratory discoveries about fresh foods and their effects on health paralleled Price's epidemiological studies. While carrying out experiments with the adrenal glands of cats, Pottenger noticed that when fed on scraps of raw meat, the animals were much healthier. They were also far less likely to die from surgery than those fed on cooked meat. So remarkable did this seem to him that he decided to conduct controlled studies to explore the phenomenon. Protocol in his experiments was carefully observed. They met the most rigorous scientific standards of the day and lasted for ten years, spanning several generations of animals.

Pottenger fed part of his colony of 900 cats on pasteurised milk, cooked meat and codliver oil. He found that these animals developed a high incidence of allergies, sickness and skeletal deformities. As one generation begat another they tended to produce smaller litters of weaker, low birth weight animals. Another group of cats Pottenger fed on the same diet but this time the meat was raw and the milk unpasteurised. These animals were healthy, had good skeletal structure and were normal in their behaviour. Their offspring were healthy through several generations, unlike the offspring of the cooked-food group which, like the people Price studied on the standard modern Western diet, showed increasingly pronounced abnormalities in physiology and behaviour. He showed that cooked foods can interfere with the normal behaviour of an animal and that the life-damaging effects of eating cooked foods are passed on from generation to generation. Finally he discovered that inherited damage induced by eating cooked foods requires four generations of animals nourished on raw foods to correct.

Pottenger then began to study the effect of nutrition on human health and to look at the effect that dietary change has on both health and disease, with a particular concern for the damage brought about by eating a diet of chemically fertilised, processed foods and proteins denatured by heat. He was not only an excellent laboratory technician but also a first rate practitioner. From his clinic in Los Angeles his reputation for astounding cures of resistant illness grew rapidly and brought him worldwide recognition. He insisted on treating his patients on a diet of raw foods which included raw vegetable juices and even a raw liver 'cocktail' which was much admired for its curative properties but which, according to many of his patients, tasted revolting!

The biogenic way of life

It is not fair to pretend, however, that the virtues of raw fruit and vegetables are a discovery of the twentieth century. In the 1920s a young French philologist named Edmond Bordeaux Szekely, researching in the archives of the Vatican, came upon the writings of the Essenes, a monastic sect that flourished on the shores of the Dead Sea at the time of Christ. These ancient texts, which undoubtedly enshrine wisdom more ancient still, give quite specific instructions about health and healing and stress the use of fasting and raw foods to achieve potent mental, physical and spiritual health. Szekely's translation of the Essene manuscripts (originally into French) was published in English in 1937 and attracted very wide attention. The French writer Romain Rolland, winner of the Nobel Prize for literature in 1915, was so fascinated by the prescriptions of the Essenes that he became co-founder, with Szekely, of the International Biogenic Society. The society is now based in Costa Rica and is still active and dedicated to researching, applying and propagating the teaching of the Essenes. Until his death in 1979 Szekely gave yearly or bi-yearly lectures all over the world to teach the 'biogenic' way of life.

Biogenic practitioners regard all illness as the manifestation of a single disease: disharmony. In this they are in accord with ancient Taoist philosophy and medical tradition which also teaches that the proper aim of humankind is the search for *wu wei*, harmony. The biogenic way of eating consists of about 40 per cent fruit – because fruit is ideal for ridding the body of its own wastes and of environmental pollutants – and 30 per cent raw vegetables. Grains, dried fruits, dairy products, seeds and nuts make up the rest. The recipe for Essene bread on page 264 uses the full goodness of raw sprouted wheat rather than cooked flour.

Healing traditions

The European tradition of using uncooked foods to heal and to promote high-level health continues to thrive at such famous clinics as the Bircher-Benner (Zurich), Ringberg (Tergensee, West Germany) and Biologisches Sanatorium (Bayern, West Germany). In Sweden Professor Henning Karstrom and many eminent colleagues continue to research and to teach about eating uncooked foods. In Finland the volume and complexity of the research done by scientists such as A. I. Virtanen and Pentti K. Hietala into the specific biochemical properties and physiological effects of raw foods is quite staggering. The Finns have a special interest in raw foods for farm animals as well as for humans. In Britain, Australia, New Zealand and South Africa there are many physicians and naturopaths who quietly heal their patients on raw diets. Even in America, where high-tech, pill-oriented medicine is strongest, the use of uncooked foods is growing – one of America's most prestigious medical corporations, Kaiser-Permanente, now boasts a Health Improvement Service which uses high-raw diets to treat conditions such as obesity, high blood pressure and diabetes. All over the world live foods and juices are part of the 'gentle' treatment of cancer.

Nevertheless the principles of healing underlying the use of raw foods are still foreign to the training most doctors receive. Also many British and American doctors are at a disadvantage when it comes to reading the scientific literature, for much of it is not in English, a reflection of the more serious attention paid to the subject elsewhere. But there is no doubt that the present health crisis in the Western world and the growing demand for a 'whole person' approach to health and illness is pushing to change all this. High-raw eating is a way of living whose time has come.

3   CAUTION: COOKING MAY DAMAGE YOUR HEALTH

No one would question that cooked foods have the ability to sustain life. What is questioned by doctors and scientists involved in research into raw diets is whether cooked foods are capable of regenerating and enhancing health. For, unless the genetic inheritance of a person is exceptionally good, a diet too high in cooked foods can lead to slow but progressive degeneration of cells and tissues, and encourage early aging and the development of degenerative diseases. Why? Some of the reasons no doubt depend on the fact that many essential nutrients are destroyed by cooking. Studies have shown that food processing and cooking – particularly at high temperatures – also bring about changes in the nature of food proteins, fats and fibre which not only render these food constituents less health-promoting to the body, but may even make them harmful. As H. Glatzel, the distinguished German nutritionist says, 'No other medium besides warmth, in its various applications, accomplishes such significant alterations of the structure and substances of raw foods.'

The prisoner of war diet

If you had been a prisoner of war in Japan in the last war you would have been fed on a diet of brown rice, vegetables and a little fruit, a diet containing a mere 729–826 calories a day per 154 lb/70 kg of body weight. Just how far this falls short of minimum recommended requirements is shown in the table below.

	Daily prisoner of war diet	Daily minimum recommended intake
Protein	22–30 g	60–70 g
Carbohydrates	164–207 g	200–400 g
Fat	7.5–8.5 g	10–11 g
Calories per 154 lb/70 kg body weight	729–826	2150

In 1950 it occurred to Dr Masanore Kuratsune, head of the Medical Department of the University of Kyushu in Japan, that the Japanese prisoner of war diet might be a stunning way to validate previous studies comparing the effects of raw and cooked foods, and the guinea pigs he chose were himself and his wife. Both followed a raw version of the diet for three different periods: 120 days in winter, 32 days in summer and 81 days in spring. During this time Mrs Kuratsune was breastfeeding a baby, and both she and her husband continued to do their usual work. Both continued in good health. In fact Mrs Kuratsune found that nursing was less of a strain than before the experiment. Then they both switched to eating the same diet in cooked form and . . . all the symptoms of the hunger disease that so devastated the inmates of the Japanese camps – oedema, vitamin deficiencies and collapse – rapidly showed themselves. They were forced to abandon the experiment. The grossly inadequate diet that had maintained their health to start with, even the health of a nursing mother, did drastic damage when it was eaten cooked.

Many thousands of laboratory animals have been experimented on to prove the same point. In India Sir Robert McCarrison fed monkeys on their usual diet, but in cooked form. All of these animals developed colitis (inflammation of the colon) and post mortems revealed gastric and intestinal ulcers too. In Switzerland O. Stiner did parallel work with guinea pigs. On a cooked diet his animals quickly succumbed to anaemia, scurvy, goitre, dental caries, degeneration of the salivary glands and, when 10cc of pasteurised milk was added to their daily diet, arthritis as well. Later Francis Pottenger reported allergies and inherited abnormalities in his colony of cats when he fed them on cooked milk and meat. From these and many other animal studies the clear picture that emerges is that diets capable of sustaining health, if eaten raw, damage health if eaten cooked.

So let us take a look at some of the important food constituents – vitamins, proteins, fats – that are most vulnerable to cooking and other forms of processing.

Precious vitamins destroyed

Vitamins – the word did not enter the dictionary until 1934 – are organic substances which the body requires in very small amounts to carry out its thousands of building-up and breaking-down operations. Some vitamins the body can manufacture itself, vitamin D for example, but the others must be taken in with our food. Vitamin C and the vitamins in the B group are water-soluble, which makes them especially vulnerable. As well as being very sensitive to heat, they leach out of food when it is soaked, blanched or boiled. Put a cabbage into cold water and bring it to the boil and you destroy 75 per cent of its vitamin C content. Cook fresh peas for just five minutes and you wipe out 20–40 per cent of their thiamin (one of the B vitamins) and 30–40 per cent of their vitamin C. Other B vitamins especially at risk in vegetables are folate, riboflavin, and inositol. Often of course the remnants of these fragile substances are thrown away with the cooking juices. Untreated milk contains 10 per cent more B vitamins (B_1, B_6 and folate) and 15 per cent more vitamin C than the heat-treated pasteurised product.

Vitamins A, D, E and K are fat soluble and so less at risk, remaining relatively stable up to about 212°F (boiling point of water). Even so up to 50 per cent of the vitamin E in food can be destroyed by frying or baking. Even the A vitamins carotene and retinol are destroyed at high temperatures.

Vital vitamins are lost in the preserving and canning process as well. The American expert in trace elements and minerals Henry A. Schroeder found, for example, that commercially frozen vegetables were seriously lacking, by as much as 47 per cent, in some of the important B vitamins (pantothenic acid and B_6) found in their fresh counterparts. Canning inflicted even greater vitamin losses – up to 77 per cent. Wheat and other grains lose between half and almost all their vitamin B_6 and between a third and three-quarters of their pantothenic acid when they are processed and refined. A large proportion of minerals and trace elements in grain is also lost during processing.

Proteins deformed

A protein is a chain of amino acids, some 20 of which are known in nature. Strung together in special sequences amino acids make up all the proteins there are. However only eight to ten of them appear to be essential for human nutrition, and our bodies need all of them almost all of the time. When proteins are heated some of these amino acids become so 'denatured' (changed in their molecular structure) as to render them useless. The digestive enzymes in the gut simply cannot process them onwards. Some amino acids are destroyed completely. If you grill steak to 239°F the amino acids cystine and lysine are lost. Glutamine, which appears to help prevent arthritis, may also be destroyed by heat. Damaging proteins by cooking is not only wasteful but also obliges one to eat more of it to get the amino acids one needs, which is risky in view of the links between high protein consumption, early aging and the development of many degenerative diseases.

About 10 per cent of the proteins in the whey (liquid) fraction of milk, higher in food value than the proteins in the curd (solid) fraction, are denatured during pasteurisation, 70 per cent during UHT processing and 75 per cent during in-bottle sterilisation. There is also evidence that cooked milk proteins, and also cooked meat, poultry and egg proteins, chemically bind with vital minerals making them unavailable for use by the body.

Another protein-related finding, of import to the health of present and future generations, comes from food toxicologist Leonard Bjeldanes and his colleagues at the University of California in Berkeley. They found that cooked eggs and beef contained substances which caused genetic mutations in the bacteria they were tested on, and the longer and hotter the cooking the greater the mutagenic activity of these substances. Frying and grilling were most detrimental, roasting and baking less so.

Enzymes, which catalyse all dismantling and construction work in the body, are also proteins and as such they too can be denatured or destroyed by heat. Some of the enzymes present in raw foods seem to be important in ensuring that we make use of other nutrients in them. We say more about enzymes in Chapter 5, but for the moment a single example will serve. Milk contains a group of enzymes called phosphatases which specialise in breaking down phosphorus-containing compounds. They are destroyed when milk is pasteurised. The result is that most of the calcium milk contains becomes insoluble, making milk constipating.

Danger: hot fat

When fats are heated to high temperatures the molecular structure of their constituent fatty acids changes. Thus altered they can be non-assimilable, poisonous, even carcinogenic. This is why it is better not to fry food at high temperatures, or reheat cooking oil, or use oil already fried in. In the heat processes involved in making margarine, cooking oil and countless convenience foods manufacturers convert valuable 'cis' fatty acids, which the body needs and can make use of, into 'trans' fatty acids, which the body cannot use. This is why it is possible to eat a lot of fat but fail to get the fatty acids you need. Recent research has suggested that a lot of people suffer from fatty acid deficiency.

Although unsaturated fats (such as corn, sunflower, safflower, soy and wheatgerm oil, and margarine) in small quantities are necessary for health and life, they become potentially poisonous when subjected to heat. Dr Rakel Kurkela at the University of Helsinki showed this very dramatically with animals in his laboratory. Some he fed with raw, unheated safflower oil, rich in unsaturated fatty acids, and others he fed with the same oil heated in the presence of oxygen, which is exactly what you do when you heat oil in a frying pan. Though he continued to feed both groups with their normal laboratory diet throughout the experiment, the first group thrived and gained weight and the second deteriorated and eventually died. Analysing heated safflower and other unsaturated oils Kurkela found they contained numerous poisonous compounds. Some of these are powerful oxidisers which bring about damaging structural changes to cell membranes, cell nuclei and proteins. Others, such as malonaldehyde, are directly cancer-inducing. If you must fry food, olive oil is probably safest since it contains only four fatty acids, but it should never be heated to smoking point.

The cooked invaders

Though there has been a strong swing away from meat and fat towards fibre and fresh foods most people in Britain and America continue to exist on a diet high in cooked and processed foods. But over a period of years such a diet deprives the body of essential nutrients and leads to sub-clinical deficiencies. With its renewal and maintenance mechanisms impaired the body loses its resistance to stress, fatigue and illness.

Research done by Paul Kouchakoff at the Institute of Clinical Chemistry in Lausanne in the 1930s throws an intriguing sidelight on the phenomenon of resistance and its relationship to cooked foods. What his work implies is that the body recognises cooked and processed foods as harmful invaders and does its best to try to wipe them out. Simply put, white blood cells (leucocytes) start rushing to the scene of the invasion (the intestines) as soon as food enters the mouth. The phenomenon is called 'digestive leucocytosis'. Until Kouchakoff's work it was thought to be a perfectly 'normal' reaction to the ingestion of all food. But Kouchakoff found that when food is eaten raw digestive leucocytosis does not occur. The number of white cells in the bloodstream did not increase when his volunteers ate raw food. Processed and cooked food, however, reliably triggered off white cell mobilisation. Interestingly leucocytosis does not occur if you eat something raw before you eat something cooked. The absence of the phenomenon when raw food is eaten may be due to certain aromatic substances in it, or possibly to its special blend of aromatic substances, enzymes, acids and natural sugars which encourage digestion and full assimilation of nutrients.

The implications of leucocytosis are these: every time white blood cells flock to the intestines to deal with cooked food the rest of the body is left undefended; continual red alerts – three or more times a day, year in year out – put considerable strain on the immune system. Raw foods leave the white blood cells free for other tasks and save the body the effort of a defensive action, thereby strengthening its resistance to disease.

4 RAW FOOD VERSUS DISEASE

The healing and health-promoting properties of uncooked foods have been demonstrated innumerable times in the biological clinics of Europe – Privatklinik Bircher-Benner in Switzerland, Dr Lars-Erik Essen's Vita Nova clinic in Sweden, Dr Josef Issel's Ringberg-Klinik in Tegernsee,

West Germany, and Klinik Prof. Werner Zabel in Bayern, West Germany, to mention only a celebrated few. Uncooked diets, coupled with other natural methods of healing such as hydrotherapy and exercise, are used to treat all kinds of illness – cancer, leukemia, arthritis, eye disorders, diverticulosis, hormone disturbances, ulcers, migraine, colds, fatigue, mental illness, heart and circulatory diseases, diabetes, stress ailments, obesity, back pain, anaemia and a hundred other common ailments.

Helping the body to heal itself

The philosophy of the biological approach is that sickness – whether it is 'caused' by viruses, germs or genetic changes – is the result of disturbances in the body's natural chemistry and that once these disturbances are corrected the body's own healing forces, which are considerable, can deal with the 'cause'. Some of the ways in which raw foods achieve this rebalancing act will become apparent as we go along.

When Professor Hans Eppinger, chief doctor at the First Medical Clinic of the University of Vienna, and his colleagues investigated why uncooked foods can successfully treat resistant illnesses such as heart disease, hypertension, kidney and blood diseases, alcoholism and arthritis, they found that they affect the body on a cellular level in many important ways. For instance they raise micro-electric potentials throughout the body. Increased electrical potentials in tissues is a direct measure of the 'aliveness' of cells. Where it occurs metabolic functions are heightened, congestion and swelling in tissues decrease, cell respiration or oxygenation increases, the body's overall resistance to illness improves and a speeding up of healing processes occurs.

The tensions of vitality

Health – indeed life itself – depends on the constant interchange of chemicals and energy between the bloodstream – which via the capillaries supplies the tissues of the body with oxygen and nutrients and carries away cellular wastes – and the cells. This interchange takes place through two thin membranes and a fine interstitial space. It occurs in a living organism only because cells and capillaries have what is known as 'selective capacity', which means they are able to attract the substances they need and reject what is harmful or unnecessary. This selective capacity is the result of antagonistic chemical and micro-electrical tensions between cells in a living system. When you die it is lost completely. The stronger the tensions – the more intense these antagonisms – the healthier and more vital your body will be.

Ill health, on the other hand, is characterised by a decrease in chemical and micro-electrical tensions and a loss in selective capacity. This in turn leads to a lowering of cell metabolism and a slowing down in cell reproduction, a weakening of the capillary walls and the gradual development of a sticky 'marsh' which builds up in the interstitial spaces from excess waste products. This marsh, or tissue sludge, encourages degeneration, favours the development of bacteria in the tissues and encourages the kind of genetic damage associated with aging. It also further lowers cell metabolism.

In this way the vicious circle of chronic illness begins. The actual appearance of symptoms may take some time to develop. In the meantime the person feels chronically fatigued and lacklustre. He or she is living in a state of 'half-health' unaware that something is wrong because as yet there are no clear disease symptoms.

At the University of Vienna scientists showed that raw food steadily increases selective capacity by heightening electrical potentials between tissue cells and capillary blood. This improves the ability of capillaries to regulate the transportation of nutrients, and gradually detoxifies the system, removing the sticky marsh that further lowers vitality. In short, a raw diet breaks through that vicious circle of disease, replacing it with a 'circle of health'.

Arthritis – away with toxins

Take arthritis for instance. Many people accept that the stiffness and pain of arthritic joints are an inevitable part of growing old. But is arthritis part of normal aging? Most experts in the use of uncooked foods insist that it is not; they view arthritis as a toxic condition which builds up as the result of bad dietary habits. A cleansing regime based on fresh uncooked foods gives the body a chance to dispel the misery-causing toxins responsible for painful joints, to improve cellular exchange and to increase cellular vitality so that gradually the condition heals. Dr Lars-Erik Essen of Sweden's Vita Nova clinic, famed for his successful treatment of arthritis, prescribed short fasts of three to five days followed by a high-raw cleansing diet. Dr Carl Otto Aly, a disciple of Are Waerland, founder of the Swedish Health Movement, uses a low-protein high-raw diet. In Britain general practitioners such as Dr Gordon Latto and Dr Phillip Kilsby have cured many resistant cases of crippling arthritis with high raw diets. They claim, as do their European colleagues, that raw regimes stimulate the body to heal itself.

Diabetes – the raw food cure

Diabetes is another widespread ailment which can be improved through raw eating. Diabetes is an illness in which the pancreas does not produce enough of the hormone insulin. Insulin works rather like a key, making cell membranes permeable to energy-giving glucose. Without enough of it, glucose accumulates in the blood and eventually overflows into the urine. As well as having to manage their illness for many years diabetics are faced with a higher risk of heart disease and cancer. Until very recently it was assumed that because high blood glucose levels are what one is trying to avoid, diabetics should not eat carbohydrates. They should eat lots of protein instead. A high-protein/low-carbohydrate diet, together with insulin injections, is the traditional method of controlling the illness. But is it the best method? A high-raw diet, low in protein and requiring no special 'diabetic' foods can, it appears, not only reduce the amount of insulin a person needs but also, in some cases, eliminate the need for it altogether.

The Schweitzer experience

The great Albert Schweitzer was a severe diabetic. When he sought the help of the raw food pioneer Max Gerson, he was very ill indeed and taking huge doses of insulin. Gerson took him off his high protein diet, commenting that since it is the pancreas that has to supply most of the enzymes needed to digest protein and since it is the pancreas which is ailing already in diabetes, why flog a dead horse? Poorly digested proteins only create more than their fair share of toxic wastes. Gerson put Schweitzer on a regime of fresh raw vegetables and lots of vegetable and fruit juices, including apple juice with all its fruit sugar. Ten days later Gerson judged it safe to reduce his patient's insulin by half. A month later Schweitzer needed no insulin at all. His diabetes never returned and he remained healthy and very active until his death in 1965 at the age of 92.

More recent evidence that diabetes yields to raw food treatments comes from Dr John Douglass, head of the Health Improvement Service at the Kaiser-Permanente Medical Center in Los Angeles. Some of his patients have been able to stop using insulin altogether, while others have reduced its use to a minimum. One of his star cases, a brittle juvenile diabetic, was weaned off insulin and eventually off oral anti-diabetic drugs as well by a 90–100 per cent raw diet. Douglass does find, however, that some diabetics need to restrict the amount of fresh fruit they eat, because fruit contains a lot of sugar. One patient who failed to respond was found to be eating 18 bananas a day!

The fibre factor

The efficacy of raw diets in diabetes is thought to be related to fibre – almost by definition a diet of raw fruits and vegetables is a high-fibre diet. As far as the diabetic is concerned the most desirable property of fibre is that it slows down absorption of glucose into the bloodstream. David Jenkins at Oxford, and others, have shown that after a high-fibre meal blood sugar does not rise as much as it does after a low-fibre meal. If the level of insulin in the blood is low, or if the patient is trying to make do with less injected insulin, slowly absorbed glucose does not swamp its limited capacity to make cell membranes permeable.

Douglass has also speculated that because fibre passes through the gut in 18–24 hours when it is consumed in large quantities (for the average Western cooked diet throughput time is 80–100 hours) there is less chance of the body being damaged by waste products in the colon. The longer wastes stay in the colon the more likely they are to decompose, producing gases which diffuse into the bloodstream and interfere with the way sugars are metabolised.

The high redox issue

Another general attribute of raw foods as valuable in treating ill health as in improving good, and which may play an important part in treating diabetes, is the very active nature of many of the substances they contain. By definition molecules that are highly active are also unstable; that is they have a high tendency to lose electrons to and acquire electrons from other molecules (a chemist would say they have 'high redox potentials'). Vitamin C has this tendency *par excellence*, but so do many other molecules – other vitamins, proteins, enzymes, fats, minerals and unknown factors – in raw food. To put it somewhat crudely, they ginger lazier molecules into action, encouraging greater energy exchange. But if food is cooked the chemical activity of many of its ingredients is reduced.

Douglass believes that the redox potential of uncooked foods – their ability to awaken relatively inert molecules – is an important factor in their potential for healing. Like vitamin C they encourage optimal electron exchange, imparting liveliness to the body's cells and systems and enhancing health. He points out that 'Optimal electron transport speed ... may not occur in denatured protein since the molecular matrix is altered. Cooking, of course, denatures protein.' Another researcher, Dr Chiu-Nan Lai, who has carried out a number of studies to determine the protective properties of chlorophyll in uncooked foods, describes it this way: 'Raw food has a higher redox potential than cooked food. Cooking destroys the oxygen-containing enzymes as well as plant tissues rendering the food more anaerobic. Putrefying bacteria which require a low redox potential environment to grow will thrive on dead tissues but not on living tissues. Raw food then is more clean.'

Raw foods versus cancer

The higher redox potentials of raw foods are probably a major reason why they form the foundation of all the 'gentle' approaches to cancer treatment and prevention. The most recent report from the United States Academy of Sciences on the relationship between diet and cancer is based on a survey of some 10000 research papers. It recommends greater emphasis on fresh fruit and vegetables in the diet. Vitamins A, C and E, which occur in good quantity in fresh leafy green vegetables and fruit, are known to discourage cancer. For example, a great deal of recent research has shown that the retinoids (forms of vitamin A) inhibit chemically induced neoplasia (new tissue growth) of the breast, bladder and skin in humans. Vitamin C, the survey cautiously comments, may 'lower the risk of cancer, particularly gastric [stomach] and oesophageal [throat] cancers'. Nevertheless in more than three years of research on mice at the Linus Pauling Institute in California a raw food diet – fresh apples, pears, tomatoes, carrots, wheat grass, sunflower seeds and bananas – was found to have cancer-preventing properties equal to those of a normal diet plus massive amounts of extra vitamin C. Even more spectacular resistance to cancer-including ultraviolet radiation was achieved when the all-raw diet was supplemented with huge doses of vitamin C. Other animal studies show that vitamin E has the power to inhibit chemically induced tumours.

Vegetable fibre has also been shown to be protective against certain kinds of cancer. And specific vegetables – Brussels sprouts, cabbage, cauliflower and broccoli – contain compounds which have been shown to lessen the effects of environmental cancer-causing agents.

It is the belief of those who treat cancer by biological methods rather than by drugs and radiation that malignancy is not something which descends out of the blue on a helpless victim, but the final stage of slow poisoning, especially of the liver, by metabolic wastes and environmental pollutants. Often this slow poisoning is the result of an unbalanced diet, a diet excessively weighted towards proteins and fats and/or refined and processed foods. An excess of protein and a deficit of vital nutrients can cause all sorts of mayhem at the cellular level – 'tired' cells are bad at picking up oxygen and nutrients and eliminating wastes, pushing the whole sodium-potassium, acid-alkaline balance of the body in a direction which breeds cancerous change.

The potassium factor

This sodium/potassium balance and good oxygenation of cells are particularly important in the prevention and treatment of cancer. Sodium and potassium work together to maintain an osmotic pressure between intracellular fluids (those within cells) and extracellular fluids (those outside cells). Potassium compounds predominate mainly in the cells of muscles, soft tissues, organs and blood vessels. Sodium is found primarily in the blood plasma and interstitial fluids. The better each predominates in its own sphere, the greater the balancing tensions between them and the more vital an organism will be.

Sodium and potassium are nutritional antagonists. When there is an excess of one the balance is disturbed and health suffers. Imbalances between sodium and potassium almost always err on the side of too much sodium and too little potassium. Indeed many people in Britain and the United States appear to suffer from some degree of potassium deficiency because of the foods they eat and the way they are cooked and processed. Organically grown foods consumed raw are high in potassium and low in sodium. In artificially fertilised foods the sodium content is higher and the potassium content lower. When food is cooked sodium is added to it in the form of salt. Processed foods are flavoured with massive amounts of salt. Excess salt, together with antibiotics and other drugs, causes sodium to be drawn into cells and potassium to move out of them as active sodium-extruding mechanisms are impaired or break down.

As the vital difference between the inner and outer environment of a cell gets less and less every single process in it begins to suffer. Unable to absorb or excrete efficiently, it ceases to carry out vital manufacturing operations, toxic wastes build up inside it and all sorts of rubbish accumulates outside. The symptoms of this clogging up and slowing down at cellular level are fatigue, lowered immunity and finally disease. Raw foods, with their high potassium content, appear to be able to throw this insidious process into reverse.

Expert in the dietary treatment of cancer Max Gerson believed that the beginnings of all chronic illness lie in this loss of potassium from the cells as a result of a gradually developing sodium-potassium imbalance in the body. This imbalance, he claimed, results in serious disturbances in the body chemistry. For not only is potassium an important nerve conductor, it acts as a catalyst for many body enzymes and is essential for proper muscle contraction, including contraction of the muscles of the heart and the muscles involved in digestion. Potassium is also vital for the conversion of glucose into glycogen in the liver. A healthy liver will contain twice as much potassium as sodium. Too little potassium causes cardiac abnormalities and can also result in high blood pressure. Low levels of potassium are associated with chronic fatigue. Potassium also has an affinity for oxygen; enough of it encourages good cell respiration or oxygenation. This is another important factor in the prevention and treatment of cancer.

Cell respiration – a key to health

The way a diet of raw foods increases cell oxygenation is just as important in the healing of a sick body as it is in protecting against illness, including cancer. In the development of most chronic illness, regardless of the specific disease, lowered cell respiration is evident. Another expert in cancer, Nobel laureate Otto Warburg, Director of the Max Planck Institute for Cell Physiology in Berlin, discovered for instance that while normal cells use oxygen-based reactions as their source of energy, cancer cells are different. They appear to derive their energy from a glucose-based chemistry instead. Other researchers, such as Heinrich Jung, and P. G. Seeger, confirmed Warburg's work and showed that cancer, like many other degenerative diseases, arises from a disturbance in cellular respiration which results not only in a lowering of energy but in a serious disturbance in metabolism in the organism as a whole. When normal cell respiration is restored by a raw diet, the vitality of the whole organism and its immunity to disease is increased.

Over quite a short time an all-raw, or nearly all-raw, diet does several things. It eliminates accumulated wastes and toxins. It restores optimal sodium/potassium and acid/ alkaline balance. It supplies and/or restores the level of nutrients essential for optimal cell function. It increases the efficiency with which cells take up oxygen, necessary for the release of energy with which to carry out their multifarious activities. With all these desirable and interactive functions to their credit it is hardly surprising that raw foods have proved effective against cancer.

Anti-cancer diets

A typical anti-cancer regime consists of organically grown food – food that has not been treated with fungicides, insecticides and artificial fertilisers (some of which have carcinogenic residues) or with additives, colorants or preservatives. Approximately 80–90 per cent of food is eaten raw and protein intake is cut down to 30g a day or less.

Protein, or rather too much protein, appears to be de-trimental to health in general and is especially implicated in cancer. Too much of it not only leads to excessive nit-rogenous wastes and deficiencies of the B vitamins niacin and B_6, calcium, magnesium and other minerals, but also puts enormous strain on the pancreas, the organ responsible for manufacturing protein-digesting and cancer-fighting enzymes. In fact many scientists consider the loss of pan-creatic function to be a major cause of cancer. A strong well-functioning pancreas is particularly good health insur-ance. Many nutritionally-oriented therapists also insist that in cancer patients it is important to 'save' most of the enzymes the pancreas produces for combating malignancy rather than digesting protein.

Anti-cancer diets are also low in fat. No more than 10–20 per cent of daily calorie intake comes from fat, all of it consumed unheated and derived directly from freshly hulled seeds, nuts and certain fruits and vegetables. Butter, margarine and processed vegetable oils are regarded with the greatest suspicion. Raw egg yolks from free-range eggs are allowed, but the only milk products used are those made from fresh raw milk – kvark, a raw unheated home-made cottage cheese, and home-made unheated yoghurt, for example. Goat's milk is thought to contain more anti-cancer and anti-arthritic factors than cow's milk.

Ferments

Fermented foods – fermented grains and juices, sauerkraut, nut and seed 'cheeses' – also play a part in most cancer treatment regimes. The lactic acid in them encourages the development of helpful gut bacteria (*Acidophilus*) which destroy their more harmful relatives (the *Escherichia coli* bacteria which meat encourages are potentially harmful) and improve digestion and assimilation. Because they are 'pre-digested', fermented foods require less effort from an already ailing digestive system. German cancer researcher Dr Johannes Kuhl, one of the first to explore the beneficial effects of lactic acid in cancer treatment, insists that as much as 50–75 per cent of the daily diet can, with benefit, be taken from naturally fermented raw foods.

Sprouted seeds and grains also figure in most cancer regimes. They cleanse the body of toxic wastes, are exceptionally high in essential vitamins, minerals and enzymes, provide easily assimilated proteins, and tend to alkalinise the blood.

The alkalinity of raw foods is a particularly powerful ally in cancer treatment too. Amongst other things it helps the pancreas to produce its cancer-fighting enzymes. According to the distinguished cancer specialist Hans Neiper the biggest challenge to the cancer healer is to find ways of breaking down the protective mucus envelope with which cancer cells surround themselves. Pancreatic enzymes have the power to destroy this mucus barrier, rendering cancer cells vulnerable to attack and liquidation by the body's immune system. Certain substances in raw fruit and vegetables (in particular the enzymes chymotrypsin, trypsin, and the bromelains, and the vitamin beta-carotene) also appear to have this power.

Live juices

Live raw fruit and vegetable juices are an essential part of anti-cancer diets. Raw juices do most of the excellent things that solid raw foods do but in a way which places the minimum strain on the digestive system. The concentrated vitamins, minerals, trace elements, enzymes, sugars and proteins they contain are absorbed into the bloodstream almost as soon as they reach the stomach and small intestine. American immunotherapist and cancer expert Dr Virginia Livingston urges her patients to drink fresh raw juices as often as possible and recommends 2 pints of carrot juice a day. Other juices she favours are apple, cabbage, cucumber, spinach, tomato and beetroot. Of course huge quantities of fresh fruit and vegetables are needed to produce juice by crushing or centrifuging. The Gerson cancer diet, for example, which prescribes ten 8 oz glasses a day of fresh carrot, apple, and green vegetable juices, uses some

1800 lb/820 kg carrots a year, 125 lb/57 kg green peppers, 145 cabbages and upwards of 1300 oranges.

Cancer now accounts for some five million of the 60 million or so human deaths in the world each year and most of these cancer deaths are in Europe and North America. Non-nutritional treatments, even the newest (heat therapy, cold therapy, laser therapy), depend on the use of external agents to kill malignant growths. They do little to help the body train its own defensive guns on the offending tissue or build up enough resistance to prevent malignancy recurring. The most recent 'miracle' drug to be tested is interferon, a commercially produced version of a substance which the body manufactures itself.

5 THE AMAZING PLANT FACTORS

Uncooked foods contain numerous substances in addition to vitamins and minerals whose effects on living organisms are just beginning to be investigated – volatile essential oils, natural antibiotics, plant hormones, pigments such as the bioflavinoids, chlorophyll and the anthocyans, and different forms of fibre. In almost all vegetable foods there are active substances which exert a positive effect on human health. But the biochemistry of plants is extremely complex and for the most part the effects of plant substances on the human body have been little studied. Some of these substances – most of which are destroyed or drastically altered by heat – appear to be particularly important for health. Various types of fibre and pigment which occur in good quantity in a diet rich in raw fresh vegetables and fruits have well-proven, but little understood, properties for encouraging high-level health. Some, such as chlorophyll, the anthocyans and pectin, even help protect the body against damage from airborne pollutants and radiation. They may also be useful in preventing cancer and in retarding aging. And for every known action of a plant-based substance there are probably a dozen yet unknown.

Non-specific resistance to illness and aging

Russian scientists such as Professor I. I. Brekhman and I. V. Dardymov of the USSR Academy of Sciences in Vladivostock have devoted decades to the study of herbs and plant foods which have the ability to increase the human body's 'non-specific resistance' to disease and aging. They have shown that certain plant-based substances which occur in the foods we eat and in the plants we use for healing not only passively affect the body in specific ways – increasing the flow of digestive juices or calming mucous membranes in the intestines, for example – but also in more general ways by strengthening the organism as a whole. Unlike pills which you can buy at the corner pharmacy, which sometimes stimulate one body system and harm another, these substances come to us in raw foods and herbal remedies in a context that is chemically balanced by nature. As such they offer a synergistic potency for high-level health. The presence of these naturally synergistic factors in uncooked foods may help explain why, as Swedish expert in raw food therapy Dr Henning Karstrom says, 'Even though you get all 50 known nutrients in your diet – i.e. vitamins, minerals, essential amino acids, fatty acids, etc. – your health will still suffer unless you also include large quantities of uncooked and unprocessed foods.' Let's take a look at just a few of these raw food plant factors and what is known about them.

Essential oils and bitters

A plant's smell may be due to as many as 50 different aromatic compounds which can be extracted as an essential oil or essence. Mint, the skin of citrus fruits, and many other strongly smelling herbs and fruits are particularly rich in essential oils. The effects of essential oils are amazingly varied. Applied to the skin some relieve irritation, because of their mild antibiotic properties. Others relieve muscular spasm and pain – until quite recently clove oil rubbed on the gums was the standard remedy for toothache. Taken by mouth others relieve coughs and sore throats, stimulate the activity of the liver and gall bladder, mildly stimulate peristalsis (rhythmic contraction of the gut), and reduce fermentation and decomposition in the gut, protecting the colon from chemicals which can form there to harm the body. Some can also be inhaled as decongestants or used to induce changes in mood and alertness, as in aroma therapy. But perhaps the most important of their effects is that they stimulate the salivary glands and intestines to secrete digestive enzymes.

Bitter factors, contained in fresh plant juices, are also principally digestive in function. They boost the secretion of digestive enzymes, exert a calming effect on the smooth muscle of the gut, and encourage better assimilation of nutrients. Plants particularly rich in bitters are mugwort, wormwood (formerly an ingredient of absinthe), angelica, sweet flag and St Benedict's herb, but they also occur in respectable amounts in many commonly eaten plants. Many apéritifs, digestifs and liqueurs contain bitters.

Plant hormones can boost immunity

Plants, like animals, depend on hormones to provide a chemical messenger service. Indeed in plants hormones take the place of a nervous system. The gibberellins are a class of plant hormones which appear to have beneficial effects on the human immune system. Another plant hormone, abscisic acid, which is plentiful in avocados, lemons, cabbage and potatoes, obligingly helps the body to use gibberellins.

So similar is the structure of some plant hormones to human hormones that it is reasonable to believe, as many researchers do, that they probably boost their actions. Secretins, another group of hormone-like substances in plants, are thought to stimulate the pancreas and the production of hormones associated with a youthful skin.

Enzymes: powerhouses for health

Perhaps the most important of the health-giving plant factors are the enzymes. They are completely destroyed by cooking.

Enzymes are the essential triggers for the metabolic machinery of every living thing from daffodil to buffalo. Some are extraordinarily powerful. The pepsin produced in your stomach breaks down the white of egg into protein sub-units called peptides in just a few minutes, but it takes 24 hours to do the same thing in a laboratory, and then only if the egg white is boiled in strong acid or alkaline solution.

There are tens of thousands of enzymes working away in the human body – some 50000 in the liver alone – breaking down food and assimilating it, building new tissue and repairing it, and manufacturing more enzymes so that the vital work can continue. An organism grows old when enough metabolic errors accumulate to injure the synthesis of its enzymes.

Many practitioners using a diet of uncooked foods for healing insist that the enzymes in raw foods are important because they help support the body's own enzyme systems. Each food, they say, contains just the enzymes and co-factors (vitamins or minerals which are linked to enzymes) needed to help break down that particular food. When we destroy these enzymes by cooking or processing, our body has to make more of its own digestive enzymes to properly digest and assimilate them. Unless you have inherited a particularly virile enzyme-replication system, without the enzymes from raw foods your body's own enzyme-producing abilities tend to wane as the years pass. By ensuring that your body has an outside supply of enzymes, these practitioners claim, you should be able to live longer, look more youthful and stay healthier.

Orthodox physicians and biochemists tend to dismiss such arguments, claiming that exogenous food enzymes are not necessary food ingredients. They say that enzymes (which are mainly protein) are no more important than any other proteins – only useful as a source of amino acids from which the body can build new proteins, and they insist that the notion of enzymes affecting health in any way is nothing more than a fantasy of ill-informed food faddists. A great deal of European research, however, indicates that they are wrong.

Professor Artturi Ilmari Virtanen, Helsinki biochemist and Nobel prize winner, showed that enzymes in uncooked foods are released in the mouth when vegetables are chewed. When these foods are crushed the enzymes come in contact with their respective substrates and they form entirely new physiologically active substances which, because of their high biological activity, are vastly important for health.

Food enzymes live on

Other European studies have shown that even the assumption that all enzymes in uncooked foods are denatured by digestion in the stomach is untrue. Extensive tests by Kaspar Tropp in Wurzburg, and by Chalaupka and others have shown that the human body has a way of protecting enzymes as food passes through the digestive tract so that between 60 and 80 per cent of them reach the colon intact. There they bring about an alteration in the intestinal flora – bacteria which live in the colon – by attracting and binding whatever oxygen is present. This removes the aerobic or oxygenated condition which is responsible for fermentation, putrefaction and intestinal toxaemia, all of which have been linked by orthodox scientists to the development of degenerative diseases including cancer. By eliminating free oxygen in the colon these enzymes help create conditions in which desirable lactic acid-forming bacteria can grow.

Protection from dysbacteria

A healthy colony of intestinal flora – the right kind of bacteria and the right quantity – produce vitamin K and apparently all the B complex vitamins. If helpful bacteria are destroyed as a result of taking antibiotics or if they are replaced by colonies of harmful bacteria then you get a condition known as dysbacteria, an 'insidious menace' to health. Dysbacteria results in the immune system being suppressed and in digestive disturbances such as wind and the formation of chemicals from bile acids; these are poisonous to the body and can lead to disease.

The importance of the right kind of intestinal flora is now being particularly stressed by scientists concerned with the prevention of cancer. A number of studies have demonstrated that diet strongly influences the enzyme activities of and the types of micro-organisms in the intestinal flora. A high-fat diet has for some time been linked with the development of cancer, probably because certain intestinal micro-organisms produce carcinogens from bile acids. When dysbacteria is present and health-damaging putrefactive bacteria are allowed to grow they can produce histamine which causes allergies. They also give off large quantities of ammonia and other chemicals which irritate the lining of the intestines and pass into the bloodstream causing toxicity of the body and predisposing it to serious illness. The enzymes in uncooked food protect against all this.

Freeze-dried plant enzymes are commonly used as nutritional supplements. When eaten with protein foods they assist with digestion and assimilation. Bromelain, from pineapple, is one of these. So is papain, from pawpaw. Chemically papain startlingly resembles pepsin, the protein-digesting enzyme produced by the stomach, and is capable of digesting 35 to 100 times its own weight in protein. Raw pawpaw has also been used to heal wounds; the papain in it digests the dead tissue which retards the healing process.

Plant fibre can save your life

But essential oils, hormones, bitters and enzymes are not the only wonder-working plant factors. Two other categories of plant substances are of vital importance: 'plantix' and plant colorants. If you have never heard of bioflavinoids or anthocyans, you are as well informed as the average general practitioner in Britain or America. They are plant colorants, as chlorophyll is. If you have never heard of plantix you have a perfectly good excuse. The word was recently coined by researchers at the Syntex Laboratories in California in order to dispel the common idea that bran is the only kind of plant fibre there is. You may be pleased to know that you need never stare another spoonful of the flaky stuff in the face again.

Plantix or plant fibre is much more than bran, which is mostly cellulose. It is also lignin (the woody fibre that keeps trees upright), pectins, gums, mucilages and hemicellulose, a relative of cellulose. These are substances which are just beginning to be seriously studied.

Fibre is what is left when all the nutrients in food have been taken out of it. But to think of it as inert, as nutritionists did for a long time, would be far from the truth. Fibre, particularly raw fibre, actively affects the gut.

Lots of plant fibre in your diet ensures at least five important things.

● More vigorous peristalsis (squeezing of food through the gut). This decreases transit time through every part of the gut, especially through the colon, and so reduces opportunities for harmful substances to damage the mucus membranes lining it.

● More bulk in all parts of the gut. This aids peristalsis and therefore transit time. It also makes you feel fuller longer, an important consideration if you are trying to cut down on snacks between meals. Bulk also ensures a stately and steady rate of nutrient absorption. By the same token harmful substances, diluted by the amount of fibre around them, get into the walls of the gut less easily.

● A low population of bacteria of the undesirable variety. These cause putrefaction of various substances in the faeces, produce cancer-causing substances from bile acids, and give off large quantities of ammonia and other chemicals which irritate the bowel lining.

● A flourishing population of beneficial micro-organisms in the gut, including those which synthesise the important B and K vitamins.

● A reduction in the amount of fat absorbed during digestion, very useful if you want to lose weight or stay the weight you are.

Pectin in particular affects the metabolism of fats and lowers cholesterol in the body. As any jam maker will know, grapefruit, oranges and apples are high in pectin. Pectin has another effect: it literally entraps molecules of heavy metals (lead, cadmium, and so on) and eliminates them from the body. A mucilaginous form of fibre called sodium alginate, found in seaweeds, appears able to prevent the absorption of radioactive strontium 90, and probably reduces the damage done by other forms of radiation too. The kind of plantix found in alfalfa is known to counteract the toxic effect of drugs, other chemicals and food additives in animals.

A diet high in uncooked plant foods provides you with many kinds of fibre, each one different, each with its own protective powers for high-level health. Such findings are of particular importance in the context of twentieth century urban life in which we are increasingly exposed to the destructive effects of the chemical pollutants and toxic substances we take in through our foods and in the air we breathe. Like radiation, these poisons encourage the kinds of damage to the body's proteins, cell membranes and specific genetic materials which are associated with aging. They also appear greatly to contribute to body toxicity, which in turn leads to the development of many common ailments from migraine to cancer. The plantix in raw foods helps protect against their damaging effects.

Green magic and the anthocyans

Among the most therapeutically exciting of all the precious compounds in plants are their pigments: chlorophyll, the anthocyans and the bioflavinoids. Living plants perform the incredible feat of converting light energy into chemical energy, a process called photosynthesis. This would not happen unless they contained chlorophyll, the pigment which gives foliage its green colour.

In 1930 Nobel prize winner Dr Hans Fischer pointed out

that chlorophyll closely resembles haemoglobin, the pigment that gives human blood its colour and oxygen-carrying capacity; the difference between the two pigments is that chlorophyll has a core of magnesium and haemoglobin a core of iron. So close is this relationship that when crude chlorophyll is fed to anaemic rabbits it restores normal red blood cell counts within 15 days and is apparently completely non-toxic. Yet chlorophyll that has been chemically refined to remove 'impurities' has no corrective effect on anaemia. On the contrary, it probably poisons the bone marrow, the site of red cell production. Spinach, cabbage and nettle juice, all rich in chlorophyll, have been used with excellent results to treat anaemia in humans too. Cabbage juice is especially good at healing stomach ulcers. In fact chlorophyll has an impressive record in the treatment of heart disease, atherosclerosis (hardened arteries), sinusitis, osteomyelitis (inflammation of the bone marrow), pyorrhea (infected and bleeding gums), and depression. It may also block the genetic changes which cancer-causing substances produce in cell nuclei, judging by research with bacteria at the University of Texas Systems Cancer Center and elsewhere. Taken internally, by mouth or rectally by enema, it curtails the activities of harmful protein-destroying bacteria and of enzymes which cause proteins to putrefy in the gut. It also makes human saliva more alkaline, an advantage if you are eating carbohydrates. For these and many other reasons chlorophyll or rather raw juices containing substantial amounts of chlorophyll are often prescribed in the treatment of allergies and malabsorption problems.

So far the anthocyans, another group of pigments, have figured mainly in the treatment of cancer and leukemia. Raw beetroot contains a particular anthocyan in large quantities – cancer patients drink the juice of just over 1 kg of beetroot daily, a little before each meal. The name most often associated with the use of beetroot juice to both cure and prevent radiation-induced cancers is that of Dr Siegmund Schmidt, a tireless anti-nuclear campaigner in the international courts.

Bioflavinoids: the amazing troubleshooters

The bioflavinoids are pigments which occur in particularly high concentration in the pith of grapefruits, oranges and tangerines, and in lesser amounts in all raw plant foods. However, they are highly active and unstable and easily destroyed by heat and exposure to air. That is why orange juice contains very little of them. To get the benefits of the bioflavinoids – and there are many – leave a little pith on citrus fruits when you peel them.

The existence of this particularly glamorous group of colorants was discovered in 1936 by Nobel laureate Albert Szent-Györgyi, the Hungarian biochemist who first isolated vitamin C. A complex of exotically named substances make up the group: hesperidin, rutin, vitamin P, flavones, flavonals, the so-called methoxylated bioflavinoids nobiletin and tangeretin, eriodictyol, and so on. Since the 1930s a great deal of research has been done in the Soviet Union, United States and Europe which has demonstrated their potent health-enhancing and health-restoring effects.

In plants themselves the bioflavinoids play a disease-preventing role. The extraordinary thing is that in humans they display the same talent. Nobiletin and another bioflavinoid closely related to it appear to have even wider anti-inflammatory powers than cortisone. Others, either on their own or in combination, actively combat infectious bacteria, viruses and fungi. Rutin, a bioflavinoid found in buckwheat, is known to lift depression. Even in relatively small doses (50mg) it significantly alters brain waves; its effect is a curious combination of sedative and stimulant, rather like that of certain substances in ginseng. Rutin is one of several bioflavinoids that also prevent superficial bruising and broken blood vessels in the skin. Nobiletin and tangeretin boost the activity of a certain group of enzymes (mixed function oxidases) which specialise in ridding the body of drugs, heavy metals and the unburnt hydrocarbons in car exhausts. Indirectly therefore these two bioflavinoids are cancer-preventing. Is this one of the reasons why so many forms of cancer yield to raw diets?

Help for sludged blood

Experiments carried out in animals and humans have demonstrated that the methoxylated bioflavinoids, especially plentiful in oranges and tangerines, significantly reduce the clumping together of red blood cells. This is not a normal tendency, but it means that blood flows less easily, clots more readily, transports less oxygen and may occasionally block tiny blood vessels, causing death of areas of tissue in vital organs. In one trial in which patients with 'sludged blood' ate three or four oranges or five tangerines a day for three weeks, blood viscosity decreased by an average of 6 per cent. More trials are in progress to find out if less fruit will achieve similar results.

The ubiquitous microbes which spread colds, flu and various low level infections also appear to quail before the methoxylated bioflavinoids. Is this one of the reasons why raw food enthusiasts appear to suffer less from these common ailments?

The wide-ranging effects of the bioflavinoids have not been anywhere near fully investigated. One of the most interesting things about them is that they seem to be most active and most useful when an organism is under the worst stress. This has led researchers to speculate that one of their main actions must be to correct the wide fluctuations in body functions which occur during illness and emergencies. In fact the bioflavinoids as a whole have such a broad range of defence actions in the body that many of the protective qualities of raw foods can probably be attributed to them.

Herbal wisdom

Almost all plants have special virtues not entirely attributable to any of the special factors we have just been discussing. Take garlic and onions, most commonly reputed to 'clear the blood'. This they certainly do – they lower blood cholesterol levels and discourage fatty deposits on the inner walls of blood vessels. They also have antibiotic properties. Garlic eaten in moderate amounts over a period of time helps to flush harmful metals from the body. Blackberries are good removers of toxic residues as well. Artichokes stimulate the liver, cabbage is anti-inflammatory and antibiotic, lettuce is mildly sedative, parsley is a kidney tonic and a natural deodoriser – the herbal knowledge of the human race has been accumulating for many thousands of years. But many of these highly active substances that give plants their gentle and extraordinary powers are damaged or destroyed by heat.

6   THE SECRET ENERGIES OF PLANTS

As we have already hinted the revitalising properties of raw foods cannot be wholly explained in terms of the vitamins, minerals and other substances they contain. Bircher-Benner believed that the vitality that raw foods give depends on their 'aliveness', which is something that defies chemical analysis. It cannot be tracked down by computing the calorific value of so many molecules of fat or carbohydrate, or by isolating and cataloguing all the different nutrients in a piece of food, or by measuring the level of those nutrients in the blood.

Living contradictions

Bircher-Benner claimed that plants contain a special form of energy directly derived from the sun during photosynthesis. When we eat plants this special energy passes into us. He sought theoretical support for his theory from physics, in particular from the Second Law of Thermodynamics.

The First Law of Thermodynamics states that the quantity of energy in the world remains constant. The Second is best formulated in terms of entropy – a measure of disorder at molecular and atomic levels. It states that energy tends to be constantly degraded from a higher to a lower order. In any system all motion eventually comes to a standstill, differences in electric or chemical potential are equalised, and temperature becomes uniform by heat conduction until finally a permanent state is reached – the whole system fades until it becomes an inert lump of matter. This state of thermodynamic equilibrium is what the physicist calls 'maximum entropy'.

But living systems are different. So long as the human body is alive it avoids decaying into this inert state of equilibrium, apparently through metabolism – eating, drinking and assimilating energy from outside.

Bircher-Benner claimed that since a very high order of the sun's energy is converted by plants through photosynthesis and then stored in them, and since the quality of this energy is degraded by all kinds of physical and chemical processes such as wilting, cooking or processing, when we eat the fresh raw plants themselves we receive the very highest order of energy possible direct from our food.

Drinking order

Some 40 years later, Austrian physicist and Nobel laureate Erwin Schrödinger confirmed Bircher-Benner's hypothesis with his own theory. But he tried to state it in terms that were acceptable to physicists: 'What is that precious something contained in our food which keeps us from death? That is easily answered. Every process, event, happening – call it what you will – in a word, everything that is going on in Nature means an increase of the entropy in the part of the world where it is going on. Thus a living organism continually increases its entropy – or as you may say, produces positive entropy – and thus tends to approach the dangerous state of maximum entropy, which is death. It can only keep aloof from it, i.e. alive, by continually drawing from its environment negative entropy . . . What an organism feeds upon is negative entropy . . . which is in itself a measure of order. Thus the device by which an organism maintains itself stationary at a fairly high level of orderliness (= a fairly low level of entropy) really consists in continually sucking orderliness from its environment.'

Bircher-Benner believed as did Schrödinger that to stay

healthy the body has to 'drink order'. We need to take into our bodies fresh living matter or foods which have the highest quality nutritive energy – energy which has not been debased by oxidation or spoiling processes or by heating.

Although many scientists are aware of Schrödinger's concept that living organisms feed on negative entropy and the idea is discussed in most textbooks on biophysics and biochemistry it is still largely ignored in orthodox nutritional teaching. Few researchers have bothered to investigate exactly how much negative entropy or what degree of orderliness exists in raw foods. One notable exception is Professor I. I. Brekhman of the Far East Scientific Centre Academy of Sciences of the USSR in Vladivostock.

The energy within structural wholes

Brekhman has coined the phrase 'structural information'. By this he means something very close to Schrödinger's 'order'. He claims that not only are the nutrients which can be measured chemically – vitamins, minerals, protein, etc. – important for health, so is the complexity of the way they and other as yet unidentified factors are combined in a particular food and the quality of energy the food carries. The processing of foods limits the structural information which they bring to an organism and thus their health-supporting properties. Fresh foods contain more structural information than cooked or processed foods. They are, in effect, more biologically active.

In experiments Brekhman has shown that foods high in structural information enable animals to carry out physical tasks for significantly longer periods than processed foods low in structural information, even when the foods compared are equal in calories and therefore, by orthodox biochemical standards, supplying an organism with the same amount of energy. Brekhman is particularly interested in certain natural pharmacological substances which appear to supply a high degree of structural information to an organism and therefore support a high level of health and energy. He quantifies the action of a plant substance or food on the body in terms of what he calls significant units of action (SUA) – a way of measuring how long an animal can continue to carry out a piece of work when fed a particular food.

American biochemist Roger Williams, famous for his discovery of the B vitamin pantothenic acid, seems to echo some of Schrödinger and Brekhman's ideas when he insists that it is time we stopped measuring the value of foods in calories alone. He says: 'In our laboratories we have recently studied an alternative criterion for judging food values. This is by measuring what we call the "trophic" or beyond-calorie value. Experimentally we ascertain how much new tissue the food in question can produce, beyond that produced in control animals where carbohydrate is supplied in place of the tested food. This method ... measures the effective presence of the entire team of nutrients necessary for tissue building and repair, including the unknown if they exist.'

Williams appears to know nothing of Brekhman's work. Although he is careful to stress that cooking and processing destroy nutrients, he tends, as do the majority of nutrition experts in Britain and the United States, to dismiss the idea that there is any special quality of energy in raw foods.

In search of new insight

If Schrödinger and the rest are right, if an organism has to 'drink order' to stay alive and the reason why raw foods are such powerful forces for health is that the structural information they carry is particularly clear and appropriate to the purposes of the living body (full health, vitality and clear awareness), then we need to ask two questions. First, what is the nature of that order? Second, in what form is it conveyed through the foods we eat – or at least, how can we measure it?

These are questions which make orthodox biochemists and nutritionists very uncomfortable indeed. For their answers are not to be found in a chemical analysis of these foods. When you enter the realm of microbiology and speak of electron transfer and the 'aliveness of cells' you may be coming closer. You do indeed point up some extraordinary properties which uncooked foods have but still you are only describing the shadows on the wall. Few scientists are comfortable with the sense that it is not the nature of reality they are delineating but merely its phantasmagoric ever-changing forms frozen for a moment in time. They mistake the shadow for the thing itself. They assume that since we know a carrot or a slice of calves' liver contains such and such nutrients and so many calories, and since we can produce those nutrients chemically in a laboratory, then by adding a little simple carbohydrate to supply the calories we can make a food that is just as good as the one we are copying. For, despite revolutionary developments in high level physics, most biochemists are still committed to the traditional atomistic notion that the universe is built out of elementary particles and that all of life can ultimately be understood by taking them apart and putting them back together. But before any scientist can begin to answer either of these questions he must first examine the underlying assumptions on which his methodologies are based and ask if these assumptions are still appropriate for what he is trying to discover.

Classical dualism is not enough

The approach to health of physicians who strongly emphasise a diet high in uncooked foods in many ways parallels that of scientists working in high level physics. Both replace the Newtonian reductionist view of reality with a quantum perspective of a dynamic universe. The Newtonian view is based on a classical dualism which not only underlies the methodologies of biochemistry and nutrition; it became an important premise on which modern medicine is based when it was formalised in Descartes's philosophy.

Descartes divided reality into two separate and independent realms – that of mind and matter – *res cogitans* and *res extensa*. This Cartesian dualism made it possible for scientists to treat matter objectively, as something completely separate from themselves to be taken apart, analysed and categorised. And indeed such a paradigm of reality has been enormously *useful*. It has made possible the isolation and control of micro-organisms behind much widespread disease in the late nineteenth and early twentieth centuries, from tuberculosis to typhoid fever and smallpox.

In the realm of biochemistry and nutrition this dualistic thinking has enabled scientists to establish the nutritional causes of diseases such as beriberi and scurvy and to isolate the 'missing' substances without which disease symptoms appear. And following such an assumption about the nature of reality, biochemists have been able to isolate, quantify and categorise the 50 or more nutrients so far known to be necessary for life.

But every dominant paradigm has its limitations. In the realm of biology and physiology such thinking has also led to the notion that the human body is little more than a machine made up of a lot of different parts which can be analysed into a collection of cause-and-effect relationships. From this world view comes the notion of disease as an outside entity – a cruel act of fate caused by some external threat like a microbe – something which we are neither responsible for nor able to heal ourselves.

Wanted: a scientific revolution
Thomas Kuhn in *The Structure of Scientific Revolutions* says that every dominant paradigm eventually expands to the limits of its methodologies and becomes no longer useful. That, we believe, is just what is happening at the moment in the field of biochemistry, nutrition and medicine. Newtonian physics and Cartesian dualism, useful though they have been in the research which led to the control of epidemic diseases and illnesses caused by gross malnutrition, are inadequate to deal with the kind of chronic illnesses which have been called the 'diseases of Western civilisation' – coronary heart disease, cancer, diabetes, arthritis, gastric ulcers, emphysema, etc.

They are also practically useless if science is to discover

ways of helping men and women to live at a state of high level health – not just free of overt symptoms of disease but feeling positively good and having a high resistance to the process of aging. For the achievement of these goals appears to involve many interrelated considerations – your relationship to stress, your psychological orientation, social and environmental factors and, perhaps most important of all, your nutrition. These influences are all too diffuse and far too complex to fit neatly into any world view based on Cartesian dualism. If we continue to try to make them fit our search for answers we will not only falter amidst masses of interesting but unconnected facts, but our efforts will also be (indeed in many ways they have already become) counter-productive.

We believe that to answer the questions we posed biochemistry, medicine and nutrition will need to expand their atomistic notions. They will need to move beyond the limitations of the Newtonian–Cartesian world view and discover a new dominant paradigm. They will need to acknowledge that the 'bits' which they continue to treat as separate entities are not descriptions of reality as it is but, as physicist David Bohm says, 'ever-changing forms of insight', which can only 'point to or indicate a reality that is implicit and not describable or specifiable in its totality'. Finally, they will have to develop an awareness – however subtle or oblique – of that implicit reality.

David Bohm is one of the world's most highly respected physicists. A protégé of Einstein, he has written the classic textbook on quantum theory used in English-speaking universities. In another of his books, *Wholeness and the Implicate Order*, he comments: 'People are led to believe that . . . fragmentation is "the way everything really is" and therefore do not look for alternatives. But it is not in fragmentation . . . that the understanding of life is to be found . . . life is the flow of that implicit order that animates the dead forms of the objective world, and as such life is both whole and never identical to any of the forms of existence. Therefore there is no use looking for it among the bits and pieces of those forms.' Modern biologists, he says, have little awareness or appreciation of the revolutionary character of modern physics. Most continue to believe that 'the whole of life and mind can ultimately be understood in more or less mechanical terms through some kind of extension of the work that has been done on the structure and function of DNA . . .' In modern physics, he continues, '. . . parts are seen to be in immediate connection . . . their dynamical relationships depend in an irreducible way on the state of the whole system (and indeed on that of broader systems in which they are contained, extending ultimately and in principle to the entire universe). Thus one is led to a new notion of unbroken wholeness which denies the classical idea [that the world can be analysed into separate independently existing parts] . . .'

To penetrate the mysteries of raw energy, biochemists and nutritionists will need to acknowledge that the healing powers implicit in raw foods are greater than the sum of their parts as measured in terms of nutrients and calories. They will also have to recognise that foods interact with the human organism as part of an *unbroken wholeness* before they can even begin to seriously tackle the issues involved.

There are scientists who work within this new dominant paradigm. They range from physicists to soil chemists concerned about how to grow stronger and more successful crops and many very important physicians. They acknowledge an interrelatedness that under the influence of Cartesian dualism and Newtonian physics we have long ignored. And in their laboratory experiments, some of them still formative, valuable new techniques and approaches for studying the raw energy phenomenon might be found.

Electromagnetism: the secret force?

Twentieth-century scientists willingly accept that animal tissue has electrical and therefore electromagnetic properties. The notion that plants, as living organisms, also possess electrical properties is less well accepted. And the idea that these properties might possibly interact with those present in the organisms that consume them is usually regarded as sheer lunacy.

Nevertheless, spurred on by Szent-Györgyi's now classic book *Introduction to Submolecular Biology* published in 1960, biomedical engineers are now testing the effects of external magnetic fields and small electric currents on broken bones and damaged nerve tissue. And to great effect. Broken bones knit faster and injured nerves regenerate. Robert O. Becker, pioneer of electromagnetic therapy at the Veterans Administration Hospital in Syracuse, New York, believes that a detailed understanding of the way in which electromagnetic forces influence living systems is likely to be the next great advance in biomedicine.

What is known about the electromagnetic properties of plants and their effects? Is electromagnetism one of the media through which we 'drink order'? There is some evidence that it might be. It was the Russian-born engineer Georges Lakhovsky who first suggested, in the 1920s, that the cells of plants and animals are microscopic oscillating circuits, in other words emitters as well as absorbers of electromagnetic energy. At about the same time an American scientist, E. J. Lund of Texas State University, discovered a way of measuring minute electrical potentials in plants and went on to demonstrate that plants do indeed generate tiny electrical currents and emit electromagnetic waves. These electrical phenomena not only change with the health of the plant concerned but also appear to direct and organise growth. Lund found, for instance, that bud formation is heralded by changes in electromagnetic radiation long before there is any detectable rise in the level of auxins, the hormones that mediate growth.

In 1936 another American scientist, surgeon George Crile who founded the Cleveland Clinic, published a fascinating book called *The Phenomenon of Life*, in which he suggested that it might be possible to diagnose illness long before the symptoms of it appear simply by monitoring the electromagnetic characteristics of the person concerned. This idea was based on the fact that physical changes in animal cells are also preceded by electromagnetic changes.

Then in the 1950s two Yale University professors, philosopher F. S. C. Northrop and doctor Harold Saxton Burr, suggested that the electromagnetic fields surrounding living organisms might be the source of the organisation that controls growth and species characteristics. To demonstrate this theory Burr began measuring what he called the 'life fields' or L-fields around seeds. He discovered that altering a single gene in the parent plant brought about significant changes in the fields of its seeds. He also found that by measuring the intensity of L-fields around seeds he could predict how healthy or unhealthy plants grown from them would be. If the seeds were subjected to chemicals or heat their fields became weaker.

Could the subtle energies which a living plant food emits be an important medium through which a living organism 'drinks order'? Could we be drawing on structural information of a high quality for human health when we eat fresh raw foods – information which in electromagnetic terms becomes distorted, decreased and less appropriate to the needs of the human organism when these foods are cooked? If so, then whatever subtle energies are given off by, say, a cooked leaf or vegetable should be markedly different from those of its raw counterpart and we should be able to devise methods for measuring those differences.

The mystery of plant auras

One possible method of making such measurements may be Kirlian photography. It is a technique which was originally discovered in Russia by an electrician and amateur photographer Semyon Kirlian and his wife Valentina. They found that they could reproduce on photographic paper, without the need for camera or lens, a remarkable luminescence that seemed to emanate from all living things but which was ordinarily invisible to human senses. One of its American practitioners, H. S. Dakin, describes how the Kirlian technique works: 'It is a technique for making photographic prints or visual observations of electrically conductive objects with no light source other than that produced by a luminous corona discharge at the surface of an object which is in a high voltage, high frequency electrical field.'

There are many problems associated with Kirlian photography and much controversy surrounding its use. For instance, scientists are still unsure about just what kind of luminescence they are picking up on their photographic plates. But the potentials of Kirlian photography are currently being investigated by no less than six official Soviet institutes using newly developed techniques many of which are applicable to nutrition or to the diagnosis and treatment of illness.

When the method is used to photograph plants and foods – comparing cooked foods with their raw counterparts or the leaf from a healthy plant with a leaf from an ailing or damaged one – researchers get consistent results. They find that the luminescent discharge or corona recorded on film from the raw healthy plant is significantly stronger, more radiant and wider than that from a cooked or damaged plant. When we examined Kirlian photography taken by British researcher Harry Oldfield comparing the corona produced by a raw carrot or cauliflower with what was produced after these vegetables had been cooked, we were surprised by the differences. The first radiated brilliant spikes of light harmoniously surrounding their shapes. The second showed only the dimmest evidence of a corona discharge.

Crystal patterns

Chromatography, like Kirlian photography, has also been used as a tool for measuring the energy radiated by living organisms, although it is more usually employed in chemistry, biology, medicine and industry as a means of analysing complex substances (the amino acids in a protein, for example) and detecting impurities.

The technique of using chromatography to measure energy differences between fresh and cooked foods, and between natural and synthetic vitamins, was developed by European chemist Ehrenfried Pfeiffer. Early in his career

Pfeiffer was asked by the German educationist and mystic Rudolph Steiner to search for a chemical reagent that would reveal what Steiner called 'the formative etheric forces in living matter'. After experimenting with different substances Pfeiffer discovered that when he added extracts of living plants to a solution of copper chloride and let it evaporate slowly it produced a pattern of crystallisation typical of the species of plant used and also indicative of the life-strength of the plant. Strong crystallisation patterns indicated health, weak ones ill health.

Later, when he had settled in the United States, Pfeiffer refined and simplified his crystallisation method, eventually using simple circles of filter paper treated with developer with a wick in the centre to absorb the liquid being tested. From the crystal patterns which formed as the filter paper dried he was able to detect differences between two apparently identical seeds, one damaged by heat or chemicals, the other not, and also accurately describe the condition and shape of the plant from which the seeds had come.

Pfeiffer always considered vitamins to be 'biological' rather than 'chemical' agents, and successfully demonstrated that there are big differences between natural vitamins and their synthetic analogues. His vitamin chromatograms are interesting, even beautiful, to look at. Commercially made vitamin C and the same vitamin in acerola cherry look very different. Synthetic vitamin A and the vitamin A in codliver oil give remarkably different crystallisation patterns. In general, man-made vitamins lack the vivid colours, strong clear patterns, radial lines and fluted edges of their natural counterparts. Similar differences appear when one compares fresh foods with cooked or processed foods. Other scientists have obtained similar results working with Pfeiffer's methods.

If chromatograms are indeed a measure of the special energy or life force – call it what you will – in food substances then one cannot but conclude that fresh foods have many times the 'aliveness' of their heated and processed counterparts. Even the addition of minute quantities of common food preservatives has been found to disrupt crystallisation patterns.

Dowsing for freshness and vitality

While Pfeiffer was working with his chromatography and various scientists in Russia and the West were looking at the differences in foods via Kirlian photography, a French engineer named André Simoneton had been using a dowsing technique learned from André Bovis, well known for the experiments he carried out on pyramids. Bovis found that he and others (Simoneton included) could tell with the use of a pendulum how fresh a food was from the power of its radiations. To measure these radiations he designed a biometer – a simple device graduated arbitrarily in centimetres to indicate 'microns' and 'angstroms' – which offered a range of measurements between zero and ten thousand angstroms. Using Bovis' version of this traditional dowsing technique Simoneton found that he would get consistent measurements of the freshness and vitality of a food. He discovered that fresh raw fruits and vegetables gave a reading of between 8000 to 10000 angstroms on his meter and would make the pendulum spin at high speed. Other foods such as cooked vegetables or pasteurised milk appeared to radiate so little energy that they did not make the pendulum spin at all.

Just what kind of wavelengths Simoneton was picking up was unknown. But because of his background in engineering he realised that the fact that something was there which could be measured consistently on a scale could make it of value to human health. Simoneton went on to measure a great variety of foods over many years and recorded his findings. He claimed he could measure both a food's vitality and its freshness. A food such as milk which measures 6500 angstroms when it is fresh loses as much as 40 per cent of its radiation after 12 hours. At the end of 24 hours a mere 10 per cent of its radiation remains. He discovered that pasteurising milk destroyed all its radiation as did tinning fruit. Cooking most vegetables rendered them 'lifeless'.

Simoneton reasoned that if foods give off various levels of whatever emanation he was measuring, so must other living systems. He began to measure wavelengths from human beings. He found that a healthy person gives off a radiance of above 6500 angstroms while the measurement of cancer patients is much lower. He also discovered that people ill with degenerative diseases such as cancer will demonstrate a wavelength below 4000 a long time before they show symptoms of the disease. In his quite fascinating book *Radiation des Aliments* Simoneton records his work and develops the hypothesis that the radiance in the foods you eat will either contribute to your body's own radiance – if they are above 6500 – or decrease it. To stay optimally healthy you need to eat mostly fresh raw fruits and vegetables, nuts and fresh fish because these foods best contribute to, rather than detract from, the body's own radiance.

More than the sum of its parts

Simoneton's work could easily be dismissed. It is based on dowsing for which there is as yet no scientific explanation. But what is significant about it, and about the work of scientists in Kirlian photography as well as Pfeiffer's chromatograms, is the underlying assumptions – the dominant paradigm – on which they have based their work. Unlike the 'atomists' who believe that truth is only to be found by pulling something to pieces and viewing it as nothing more than a collection of 'bits', they postulate the existence of an energetic principle which underlies life and health and which suggests to them a subtle but pervasive interconnection between mind, body and environment. Pfeiffer insisted that, as Goethe said, 'The whole is more than the sum of its parts.' In a booklet which he wrote before his death he says: 'One can ... take a seed, analyse it for protein, carbohydrates, fats, minerals, moisture and vitamins, but all this will not tell its genetic background or biological value ... a natural organism or entity contains factors which cannot be recognised or demonstrated if one takes the original organism apart and determines its component parts by way of analysis.'

Until these factors are taken into account, explanations about why raw foods have such extraordinary capacities to promote high level health will, at best, remain frustratingly inadequate. Once they are, research into the whys and wherefores of healing with uncooked foods could lead to the expansion of other areas of consciousness not even directly related to nutrition or health. In the meantime, used wisely, these simple foods – fresh fruits and vegetables from the greengrocer or the garden, home grown sprouted seeds or grains, natural unheated dairy products, nuts and seeds – which are available to everyone and can be chosen for their very low cost if necessary, might alleviate a great deal of human suffering and make possible the good use of a lot of dormant human creative energies. You don't need to understand the theories of raw energy to apply its practical benefits to your life.

TWO

THE
RAW ENERGY
EXPERIENCE

7 ENERGY, ENDURANCE AND ATHLETICS

A high-raw diet – one in which at least 75 per cent of your foods are uncooked – creates exceptionally high levels of energy and stamina. We find, as others have, that the harder we work the greater the percentage of raw food we need to eat in order to function at top peak mentally and physically and still have energy left over for play. We notice the energising effects of a high-raw diet most of all when it comes to stamina and endurance. We belong to a family of amateur runners. Unless we have been out for a run along the cliffs of Pembrokeshire or around the Outer Circle of London's Regents Park we feel something is wrong somewhere. When we began to eat a lot of our food raw we noticed that we could run for far longer without fatigue. We also had a feeling of lightness on the road, a feeling which regularly disappeared when we had been to a party the night before and graciously eaten more 'normal' food.

But it was not only physical energy that soared when we changed to a high-raw diet. We also found we could write and research efficiently for much longer periods – seven or eight hours rather than three or four – always provided we stuck to our sprouted salads and fresh juices. We also needed less sleep. And when we did sleep, our sleep was deeper and more restful. One of us (Leslie) found she needed only five or six hours' sleep a night rather than her previous eight, and felt better than ever before on it. We got up each morning bright-eyed and eager to face the new day – no crawling out of bed longing for the moment when we could crawl back in.

My dear, you need something substantial!
But we worried. We had both been brought up to believe that lots of protein (which to us meant meat) keeps you strong and healthy.

We were eating a fraction of the protein we had eaten before. Were all the women's magazines right? Would we lose our glossy hair and long, strong nails? We felt uncomfortable. Were we nourishing ourselves properly? Every time we got up from the table we felt light, not uncomfortably full as we used to. Our friends were not much help. When they saw what we were eating – fruit for breakfast, and salads piled high with home-sprouted seeds and grains for lunch and dinner, plus the occasional piece of fish or poultry – they got more worried than we were. 'My dear', they would say 'are you sure you know what you're doing? Salads are all very well but don't you need something more substantial, something that will stick to your ribs?' Their concern was valuable, although it was unsettling. It made us wonder about the wisdom of our new way of eating. Had other active people, athletes say, ever tried to live on such a regime? And if so, how did it affect their physical strength, endurance and performance? How much protein does one really need to stay healthy?

Raw energy athletics

George Allen, who held the world record for walking from Land's End in Cornwall to John O'Groats at the far northern tip of Scotland, and who kept going day and night to achieve it, lived on raw vegetables. As a youth he was epileptic and in search of a cure began to eat raw turnips and then other raw foods. His record was beaten many years later by a woman, Barbara Moore, also a raw food enthusiast. In France the celebrated herbalist Maurice Messègue trained the famous racing cyclists Fausto Coppi and Luis Ocana to victory on a strict routine of raw fruits and vegetables, whole grains and honey. American comedian Dick Gregory became a competent athlete on a regime of raw foods and juices, and occasional fasting. In 1974 he ran 900 miles/1450km on fruit juice alone. In Sweden top athletes improve their performances by following the principles set down by Are Waerland and eating a high-raw/low-protein diet. Even in the United States and the Soviet

Union where the 'eat meat to win' myth dies hardest, experts in sports medicine and athletes themselves are beginning to extol the advantages of a vegetarian regime high in raw foods. This is as true of professional as amateur athletes, whether they are monolithic weightlifters or skin-and-bone marathoners.

Until ten years ago the notion that carrots and apples rather than steak and eggs offer the greatest potential for energy, endurance and world-beating performance was pooh-poohed by nine out of every ten nutritionists and athletes. Now the vegetarian approach is being taken very seriously. The latest recommendations for athletic diets centre around complex unrefined carbohydrates and much less protein than even five years ago.

The high protein myth

Looking through the scientific literature we discovered that as early as 1866 two German physiologists, Pettenkofen and Voit, had done some experiments which showed that protein does not bring the quick energy for muscles that most sportsmen claim it does. Then, in the early years of this century, two American physiologists, Irving Fischer and Russell Chittenden, professor of physiology at Yale University, decided to find out what a diet must consist of to give the maximum physical energy and stamina. Fischer worked with a group of athletes, varying their diets and recording his results. Without exception his subjects found that they achieved the greatest fitness on a high-raw vegetable diet. They also found they needed to eat less. In his studies Chittenden fed three different groups – athletes, moderately active workers and university professors – on a diet which he made as varied as possible and which included meat. His main interest was to find out to what extent protein was essential for energy, health and endurance, and so he varied the quantity of animal foods his three groups were allowed. His results not only proved that athletes can train for and win international championships on a low-protein diet (approximately 50g a day instead of the 120g then believed best, and still believed best by many today) but also that fitness in all three groups improved when they ate fewer calories and only half the protein.

Maximum strength, capacity and endurance

Intrigued by this Chittenden decided to carry things further by experimenting on himself. Although he had no intention of changing to a vegetarian diet or of increasing his consumption of raw foods, his desire to find the lower limits of protein requirement obliged him to exclude all flesh foods from his regime. When he did he discovered that his energy levels continued to rise. A rheumatic knee joint that had bothered him for 18 months stopped bothering him. The digestive disturbances and headaches he periodically suffered from also vanished. And all on a mere 1600 calories a day and just 33.73 g of protein.

Eighty years ago then scientists had very convincingly demonstrated that eating lots and lots of protein does *not* improve athletic performance. For thirty years Jean Mayer, highly respected former Harvard nutritionist, extolled to his students the value of a diet high in complex carbohydrates as the best fuel for muscular exertion. But still Harvard coaches took their teams out to heavy steak dinners before each competition. The great protein myth dies hard.

Eimer's all-raw athletes

When we searched the scientific literature further we discovered some even more interesting research with athletes. In the 1930s Professor Karl Eimer, director of the First Medical Clinic at the University of Vienna, went further than Chittenden. He put top athletes into a high-intensity physical training programme for a fortnight then suddenly, and without any attempt at slow transition, changed their diet to one of entirely raw foods.

Protein intake dropped from at least 100g a day to only half that amount. Eimer expected a complete breakdown in athletic performance, but he was disappointed. His athletes grew stronger, faster and more supple.

So the experience of improved stamina and energy levels was not unique to us. Many others, some of them super-fit even before making the switch to a high-raw diet, had made the same discovery. When we asked physicians and biochemists why raw foods have such an energising effect they told us it was probably due to several factors.

● A high-raw diet offers, in perfect and complementary combination, all the nutrients essential for maximum vitality at the whole body and at the cellular level.

● Raw foods cleanse the body of stored wastes and toxins which interfere with the proper functioning of cells and organs and lower energy levels.

● Raw foods increase the micro-electric potential of cells, improving your body's use of oxygen so that both muscles and brain are energised.

● Since a high raw diet is almost inevitably low in protein and low in fat it makes for better functioning overall. Too much fat slowly starves tissues of oxygen and too much protein exhausts the body's mineral reserves and creates an inordinate amount of toxic waste.

Oxygen and potassium

Athletic performance is largely determined by how efficiently your body uses oxygen. This in turn depends on how good your heart is at delivering oxygenated blood to your muscles and how good your muscles are at extracting that oxygen once it arrives. Regular training helps a lot. It strengthens your heart so that it pumps more blood with each beat. It increases the number of oxygen-carrying red cells in your blood. In addition it enlarges the smaller blood vessels so that more blood can flow through them. And it also speeds up the rate at which the enzymes in muscle cells use the oxygen offered to them.

A high-raw diet specially affects the last part of the process, the absorption and use of oxygen in muscle cells. In 1938 Professor Hans Eppinger, chief doctor at the First Medical Clinic of the University of Vienna, showed that raw foods increase cellular respiration. A high-raw diet also stimulates muscle cells to absorb nutrients and excrete wastes efficiently. And in time it flushes away the noxious 'marsh' that develops between cells when too much protein is being eaten; once the marsh has been cleared, speedy oxygen, nutrient and waste exchange can be resumed. Such a diet gradually detoxifies the whole body, giving the muscle cells ideal conditions in which to produce energy. A high-raw or an all-raw diet also happens to be an excellent way for athletes to eat in the 48 hours leading up to major events.

We also wondered whether the high potassium content of raw vegetables and fruit had anything specific to do with good muscle tone and stamina. We found that potassium deficiency is a common hazard in long-distance runners; the mineral is rapidly depleted during hours of sustained exertion. If it is not replaced, even a trained athlete becomes chronically fatigued. Marathon runner and co-author of *The Sportsmedicine Book* Dr Gabe Mirkin says: 'When you lack potassium, you feel tired, weak and irritable. Could some athletes, amateur and professional alike, not be using their full potentials in performance because they, like a majority of people living on the average British or American diet, are deficient in potassium – one of the minerals most essential to good muscular activity?'

Help for aches and pains

Given half a chance most runners will tell you all about their current agonies – an Achilles tendon gone wrong, a strained muscle in the groin, a knee that gives the most appalling and inexplicable pain at unexpected moments. We were no different. In fact, the most serious athlete in our family is 24-year-old Branton. He is a keen fell runner. Thin as a rail, he bounds along cliff paths like a young stag. But Branton has always suffered from more muscle aches and joint pains than the rest of us. Sometimes it was a foot (X-rays showed no sign of anything mechanically wrong), sometimes an unpleasant ache that seemed to stretch down the left side of his body. At times his aches and pains were so bad that even he would stop exercising for a day or two. And no amount of massage, osteopathic treatment or changing of shoes seemed to help. When the two of us noticed that on a diet of mainly uncooked foods our own niggling muscle and joint problems disappeared we suggested that he might try eating more raw food to see if it helped him. After three or four months during which he gradually ate more and more of his food raw his aches and pains became fewer and farther between. In the end they all but disappeared.

Exactly why a high-raw diet achieved this no one could tell us. Perhaps it gets rid of whatever deposits in muscles and joints make movement painful. That must certainly be one of the reasons why raw diets are so helpful in arthritis.

In fact a quite frequent comment from people who make the switch to a high-raw diet is that it relieves persistent back pain. Whatever the reason, Branton revels in the pain-free movement which his new way of eating has brought him. It has made him our most enthusiastic advocate of high-raw diets for athletics. On occasion he still turns to us in amazement and says: 'I never thought it could make such a difference.' Of course not. Neither did we. Not until we made the discovery for ourselves.

8 REJUVENATION AND LONGEVITY

Can a high-raw diet keep you looking and feeling young? Can it reverse some of the age-related changes that have already taken place in your body? Does it have powers for rejuvenation? These are the questions we are most often asked by people curious about our eating habits. Although few studies have concentrated on the rejuvenating or life-extending properties of raw foods there are strong indications, both from the mental and physical state of people who live on them and from our knowledge of their biochemical, physiological and energising effects, that the answer to all three questions is 'yes'.

The aging process

Despite enormous research efforts in the past thirty or forty years we still know very little about the specific processes of aging or how to retard them. We know that two systems seem to be involved: the genetic code carried by the DNA in cell nuclei, and the immune system, with its ability to recognise 'self' from 'other'. These two systems are inter-connected in very complex ways, but together they ensure that cells renew themselves identically. They also fight off illness, neutralise toxic substances in the body, and repair damaged cells and tissues. When both systems are working properly little aging takes place.

But the slow, steady build-up of minor biochemical imbalances in the wake of stress, poor nutrition and assaults from toxic substances and radiation in the environment progressively interferes with their intricate working. Then the body begins to deteriorate. To retard the aging process any diet, treatment or regime must be able to prevent damage to these two systems and also (at least to some extent) repair damage already done.

Many animal experiments have been done to see if diet can retard aging and extend lifespan. So far the most promising anti-aging substances found in food are the anti-oxidants, nutrients such as vitamin C, E, A and some of the B vitamins, the trace element selenium, the sulphur-containing amino-acids, and certain food preservatives such as BHT. These substances help to prevent the destructive oxidation processes which break down a cell's genetic material and cause the cross-linking of molecules that result in poor replication, in other words aging on a cellular level. A number of the world's leading age researchers believe that adding some or all of these anti-oxidants to our diet in sufficient quantities can hold back many age-related changes and perhaps – though this is still very controversial – increase longevity too.

Raw foods versus aging

That a high-raw diet bolsters the immune system is pretty certain from research into the way it prevents such phenomena as digestive leucocytosis and colonisation of the gut by harmful bacteria. The protection it gives against degeneration and acute disease conditions in clinical terms alone is another strong indication that it strengthens the body's immune responses.

Various cancer studies, such as Lai's at the University of Texas which showed that wheat grass inhibits mutations in DNA and work at the Linus Pauling Institute which showed that a raw diet lowered radiation-induced cancer in mice, support the thesis that raw foods sharpen up an organism's ability to distinguish 'self' from 'other' and destroy the 'other'. Raw foods probably contain other anti-aging factors besides the anti-oxidants we know about. We know, for example, that there is a special enzyme called superoxide dismutase (SOD for short) which discourages the formation of 'rogue' molecules called superoxides and free radicals which do serious oxidative damage to every part of the body.

The anti-aging enzyme

SOD occurs quite naturally in every single cell in the body. So far four different forms of it have been identified, three of which play protective roles. Although research into SOD is still in its early stages it appears to play an important part in preventing cancer and protecting the body against radiation. Intravenous injections of SOD are now on clinical trial in patients who suffer from arthritis, muscular dystrophy, cancer and radiation poisoning. Researchers are highly enthusiastic about its ability to protect cell DNA and other body systems, especially the immune system, from the damage associated with aging. Nutritional supplements of SOD are available from health food stores. The theory is that they boost the body's own production of SOD, although there is no scientific proof so far that this is what actually happens (remember the insistence of orthodox biochemists that enzymes taken orally do not survive the digestive process?). Nevertheless many scientists insist that SOD supplements do dramatically improve the health of those who take them.

Raw foods are very rich in SOD. It and the other enzymes present in raw foods get to work on their respective substrates as soon as food is chewed, quickly forming other active compounds which are important for health.

The proof of the pudding

The changes which take place when you eat a high-raw diet speak for themselves. Skin loses its slackness and puffiness and seems to cling to the bones better. The true shape of the face emerges where once it was obscured by excess water retention and poor circulation. Lines become softer. Eyes take on the clarity and brightness one usually associates with children or with super-fit athletes.

Ann Wigmore is one of the West's greatest advertisements for uncooked foods. She is the director of the Hippocrates Health Institute in Boston, a non-profit-making organisation which teaches people to grow and prepare vegetable foods for maximum health. Wigmore, who is now in her seventies but looks much younger, lives on a completely raw diet and delights in telling the story of her own rejuvenation. At the age of fifty she was ill, chronically fatigued and very old looking. She began to experiment with raw foods and changed the way she was living. Within quite a short time her fatigue and illness vanished. Her greying hair began to return to its original dark colour. Her skin tightened as though she had had a face lift. She felt better and more energetic than she had ever felt in her life. Though many more conservative advocates of a high-raw diet criticise her passion for wheat grass and what they consider to be her rather rigid conception of what a good diet is, there is no denying that her regime has rejuvenated her and thousands of others who have sought her advice.

The rejuvenation that can be achieved on a high-raw diet is not superficial, not just skin deep. It happens at the physiological and biochemical level as well. Most of the tests used to assess age-related change – serum cholesterol, serum lipids, blood pressure – reveal changes for the better. In the diet-oriented clinics of Europe a high raw diet has been shown capable of healing many of the degenerative diseases associated with aging. Impotence and other sexual dysfunctions, which tend to increase with age, right themselves. Flagging sexual interest rekindles. Raw diets are even said to reduce the severity of senile dementia.

The perfect diet for longevity

Diets based on fresh raw vegetables, fruits, seeds, nuts and perhaps a few dairy products and the occasional piece of fish or meat, are by definition low in calories and low in protein. The importance of both these lows cannot be overemphasised. When you eat most of your food raw you do not need so many calories. Nor do you want to eat as much as you would on a 'normal' diet. Partly this is because a high-raw diet contains such a lot of fibre, and partly because it does not overstimulate the digestive system and make you want to eat more and more. Age research with animals has conclusively shown that low calorie diets can extend 'normal' lifespans by as much as 300 per cent as well as retard the physical manifestations of aging. Low protein diets discourage the build up of toxic wastes in the connective tissues.

One thing that most age researchers unreservedly agree upon is that laboratory animals live longer and healthier lives if their diet is rich in essential nutrients – vitamins, minerals, fibre, essential fatty acids and so on – but low in calories. Studies done in 1935 by Clive McCay at Cornell University showed that when animals are under-fed from weaning – that is when their calorie intake is deliberately restricted although they are given every nutrient essential for health and growth – their lifespan increases significantly. When McCay's research was first published the reactions of the scientific community were mixed. It was exciting to have proof that it was possible, and so simple, to increase an animal's lifespan, but was it ethical to try to increase the human lifespan in the same way? It would not be right to take human infants and restrict their calorie intake. After all a small percentage of experimental animals, between 2 and 5 per cent, die when subjected to early calorie restrictions. Also, although limited calories during the rapid growth stage produce perfectly healthy, if somewhat small, animals they might well jeopardise brain development in the human infant. So for a long time the implications of McCay's work for the human lifespan were pushed under the carpet.

Life extension in middle age

Then, in 1979, two age researchers at the University of California at Los Angeles decided to find out if they could extend the life of 'middle-aged' animals by feeding them a low calorie but highly nutritious diet. In experiments which have now been repeated by others, Roy Walford and Richard Weindruch showed that adult animals reach significantly greater ages if they are consistently underfed. Some of their mice lived 40 per cent longer than normal, and some of their fish to three times the normal age for their species. They also noticed that degenerative illnesses such as cancer, kidney disease and heart disease were all less frequent in animals living on a calorie-restricted diet than in control animals fed on the same diet but allowed to eat as much as they wanted. For example, 50 per cent of Walford's control group of mice ultimately developed cancer compared with only 13 per cent of the calorie-restricted group. Australian research along the same lines showed even greater divergence in susceptibility to cancer: 65 per cent of control animals developed cancer compared with only 15 per cent of the underfed animals. Underfeeding also means that when degenerative disease does develop it does so later in life. Walford's underfed mice also showed less discoloration and matting of hair and dryness of the skin than their controls. Clearly their immune system thrived on a low intake of calories.

There is also evidence that underfed animals stay physiologically younger for longer. Blood cholesterol levels go up in normal groups but not in underfed groups. Changes in the contractility of heated collagen (one of the most important clues to the rate at which an organism is aging) also occurs significantly later and more slowly in underfed groups. Extrapolating these findings to humans, Walford and others believe that a calorie-restricted but highly nutritious diet, begun even in middle age, is likely to lead to longer life and a healthier old age. Walford now follows a low-calorie regime himself, and so do some of his colleagues. He fasts on water two days a week and eats highly nutritious but very low calorie meals the other five.

Long-lived cultures

Virtually all long-lived cultures – the Hunzakuts, the Georgians, the East Indian Todas and the Yucatan Indians – live on a low calorie diet, a diet rich in fresh uncooked foods. So did the 'primitive diet' cultures that Price visited in the 1920s and 1930s and in whom he found very low incidences of degenerative illnesses. Is a high raw diet, then, an optimal diet for promoting human longevity? We believe it is. But there is another important characteristic which the diets of long-lived people have in common, a characteristic that scientists using calorie restriction to extend animal lifespans appear not to have considered: all of their diets are low in protein. This, plus the fact that the peoples who have the lowest life expectancy (average life expectancy among the Laplanders, Greenlanders and Eskimos is between 30 and 40 years) live on a diet high in animal protein, suggest that any diet designed to retard aging should contain very little protein.

The protein menace

So insidious and destructive are the effects of a high protein diet, and so extensive is the research which proves as much, that it is difficult to understand why the 'lots of protein is good for you' myth still survives. Excess protein is so damningly implicated in premature aging that it is hard to understand how anyone who is serious about caring for themselves in the long term can continue to eat large quantities of high protein foods. A diet which supplies more protein than the body needs actually causes deficiencies of many essential vitamins, including the B vitamins B_6 and niacin. It also causes important minerals such as calcium, iron, zinc, phosphorus and magnesium to leach out of the body. And during protein breakdown complex by-products are formed some of which, ammonia for example, are highly toxic. These toxic residues deposit themselves throughout the body, predisposing it to degenerative illnesses – to arthritis, atherosclerosis, heart disease, cancer, even to schizophrenia. Lots of protein certainly brings about early and rapid growth, but it also brings about early and rapid aging and disease.

One of the most dangerous protein by-products is a fatty-waxy deposit called amyloid, found in large quantities in the tissues of dedicated overeaters of meat. Dr P. Schwartz, professor of physiological pathology at Frankfurt University and a world expert on amyloid deposits and their implications, refers to amyloid as 'the most important and perhaps decisive cause of decline with age'. It not only stifles proper cell metabolism by interfering with the movement of oxygen and nutrients but also damages cell membranes and DNA.

How much protein is enough? This is a question that provokes heated arguments, even in scientific circles. The idea that lots of protein was a good thing developed around the turn of the century when Voit and Rubner recommended 120–160g a day. Then in 1905 Chittenden showed that high level health and peak athletic performances could be achieved on less than half that amount. Studies done in 1969 by Hipsley and Oomen suggested that good health is possible on as little as 15–20g a day. Ralph Bircher, who has studied the conundrum thoroughly for 50 years, prefers 50–60g which, he says, is probably on the high side and 'allows a good margin for error'. In Britain and the United States, recommendations usually hover somewhere be- tween 70 and 100g a day, but an increasing number of nutritionists now insist this is excessively high. Expert on the relationship between diet and aging, Dr Myron Winick of the Institute of Human Nutrition at Columbia University School of Medicine, has stated that for maximum protection against aging and degenerative disease the 'recommended daily intake for healthy adult men and women of almost any age is 56 g and 46 g respectively'.

Anyone who is eating a high-raw diet, however, is likely to need less protein than someone on a cooked diet. For it is not just quantity of protein that matters, but quality. And the quality of protein in a mixed vegetable diet, which supplies all eight essential amino acids in easily assimilable form, is very high indeed. The long-standing notion that you have to eat meat or eggs to get enough protein is simply untrue. As Carl Pfeiffer, author of *Total Nutrition*, says: 'There is a general belief that only meat gives us protein and that vegetables could never be equal. This is quite erroneous. It is a matter of calorie density: a helping of broccoli has a very high protein value in proportion to the calories consumed. There is also the advantage that the calories are in a high fibre form.'

Action time

Considering everything that is known about aging and the ways in which diet can hasten or retard it, it seems to us foolish not to take action now. A high-raw diet is low in calories, low in protein, but super-nutritious. And just in case you think you might have difficulty reducing your calorie intake to somewhere between 1 600 and 2 000 a day (which is what Walford and his colleagues recommend) let us reassure you that after a few weeks of high-raw eating you will quite naturally tend to eat less. You will feel fuller on raw foods because they contain a lot of fibre and because they give you a high complement of essential nutrients. Your body will not constantly 'cry out for more'. Those awful cravings which lead to excessive eating and obesity belong to cooked diets, not to mostly raw diets.

9 RAW POWER FOR SLIMMING

Weight control is easy on raw foods, so easy in fact that anyone who has a habit of going on and off slimming diets and achieving only temporary 'success' will rejoice to know that on an all-raw diet most people lose weight steadily without ever counting calories. This is what centenarian Dr Norman Walker, one of America's experts on the use of raw foods for healing, has to say on the subject: 'I can truthfully say that, without exception, every person I have ever known during the past thirty-five or forty years who has gone on such a program [an all-raw diet] has not only been able to overcome weight problems and ailments resulting from the neglect of the body, but has been able to prevent worse calamities, even when surgery was recommended.' His comment is echoed by his fellow American Dr John Douglass, a tireless experimenter with raw foods in the last fifteen years: 'For many years I struggled with obesity and was frustrated in treating patients because nothing ever seemed to work – not biofeedback or hypnosis or diets or anything. Then I discovered the potential of uncooked foods and found that the more uncooked foods patients used, the less they wanted to eat. These foods are more satisfying for patients and they lose weight on them.'

Ordinary slimming diets are notoriously poor nutritionally. Going on and off them as many slimmers do creates sub-clinical deficiencies that lead to illness and fatigue and also to the 'hidden hunger' that triggers bingeing, bad eating habits and more weight gain. Raw eating is different. Slimming on a raw diet seems to curb the craving for food which causes people to gain weight. And as it improves nutritional status so it encourages steady and continuous loss of those excess pounds.

Fat can be fatal

It is common knowledge as well as scientific fact that being fat predisposes you to serious illness – to atherosclerosis, heart disease, high blood pressure, diabetes and cancer, among many others. It is also scientific fact that the fatter you get the less efficiently your immune system protects you against infection and early aging. Raw food experts regard fat as a symptom of illness, illness directly due to poor nutrition. Correct the illness slowly through raw eating, they say, and weight will gradually adjust itself. Conventional medicine of course prefers to regard fat as a failure of will-power.

Why do so many people overeat when they eat cooked foods, especially junk foods? Because their diet is short of vitamins, minerals and other essential nutrients. The body, in a desperate bid for nutritional satisfaction, demands more and more calories. Overeating produces more poisonous wastes than the body can get rid of and so it stows them away in the layer of fat under the skin where they can do as little damage as possible. Dr Kristine Nolfi, founder of Humlegaarden, Denmark's leading raw food health centre, used to refer to excess fatty tissue in the body as 'a poison dépôt in an over-acid organism'.

Digestive overload

An overweight body is not only a deformed body with its bloated flesh, flabby muscles and knotty deposits beneath the skin, it is also a body with an inherited tendency to store fat. And it is a nutritionally starved body despite the number of calories it has consumed over the years. Its endocrine system, circulation, bones and nerves are under constant stress. As a result of overeating, eating too many nutritionally 'empty' foods or depleted foods or of eating too few nutritionally potent ones, the fat person's digestive system has become overloaded. His digestive juices are in overproduction. This leads to deficiencies in important enzymes needed to break down his foods fully and provide nutrients for cell use. So regardless of how much or how little he puts into his mouth, his cells are craving to be fed and he is plagued by constant appetite. The constant stimulation of the digestive organs can result in excess acidity of the stomach and longstanding inflammation of the intestines which further increases the craving for more food. An overstressed enzymic system can lead to food allergies and result in sub-clinical vitamin and mineral deficiencies which in turn reinforce the vicious circle of hidden hunger. The overweight person knows the pattern only too well – that feeling that you must eat and eat because nothing seems to satisfy you, but when you do you get even more hungry. Medicine calls it dysphagia. To the fat person it's just plain misery.

Diets don't work

Fat cells are much less active than other cells. They burn a lot less energy than muscle cells, for example. This means that the more fat you have relative to muscle the lower your metabolic rate is. A fat person uses up fewer calories per kilo of body weight than a normal weight person, which is why he or she can eat very little and still not lose weight. Cracking down on the calories has much less effect if you are fat.

The latest theories about why fat people stay fat concern what is called the 'fatpoint'. This is not a fixed level but the usual level of fat your body is accustomed to. Whatever weight your body has maintained for, say, a year or so, seems to determine your fatpoint. Your body maintains this usual level of fat by checking up hormonally on its fat reserves. The hormones that do this work via the brain, and the level of these hormones in the bloodstream is directly proportional to the amount of fat stored.

It is this fatpoint, and the cravings that are a sign of malnutrition, which spur the familiar phenomenon of rebound eating, uncontrolled eating as soon as you stop dieting. Rebound eating may raise your fatpoint higher than it was before you began dieting because you have alerted your hormones to the fact that your body's reserves of fat are falling. And so you put on even more weight. This is how most people's weight creeps up over the years. When you diet repeatedly you are fighting one of your body's most efficient self-protection mechanisms, and you are the one who loses.

The solution is to work with, not against, that protection mechanism. You have gradually to readjust your fatpoint so that lost weight stays lost. This means doing three things.

● Eliminating craving by giving your body all the nutrients it needs (at least 50 essential nutrients are known but fresh raw foods probably contain many more).

● Calming an irritated and overactive digestive system so that its functions return to normal. Then you will derive full benefit from the food you eat and eliminate the ravenous hunger.

● Getting rid of accumulated toxins and amyloid deposits, because these block proper assimilation and adversely affect the endocrine and nervous system.

A diet of well chosen raw foods will do all these things and it will do them gradually, without your having to pay attention to calories and without alerting your body's fatpoint defences. And the wonderful thing about losing weight the high-raw way is that you do not end up looking drawn or flabby. Skin and muscles become firm and the whole body undergoes a slow process of rejuvenation which is little short of miraculous.

Another problem familiar to dieters is irritability. Flare-ups of temper and rapid changes of mood are often due to an over-acid system. If you have a lot of fat in your blood, which is usually the case if you are dieting and getting rid of stored fat, your blood will be acidic. Unfortunately most of the foods that make up standard slimming diets also tend to increase acidity. The more acidic your system the more irritable you feel. Uncooked foods, particularly fresh fruit and vegetables, have a counteractive alkalinising effect. This means that you feel calmer, more resilient and less tired while you are losing weight.

Raw fibre fights fat

Much has been made recently of the fat-fighting virtues of fibre. We are told we should sprinkle bran on our breakfast cereals and eat baked beans on wholegrain toast. Nevertheless the fibre in raw foods is better for slimmers than the soggy, compressed fibre in cooked and processed foods. Unlike the proverbial baked bean, raw food needs lots of chewing and swallowing, an important factor in controlling the amount you eat. Raw fibre also gives your stomach the bulk you associate with feeling full, and that too helps you to control your appetite. And raw fibre travels through the colon faster than cooked fibre, which means that you will not suffer from the 'blocked up feeling' overweight people and slimmers know so well.

Cooked foods tend to leave a slimy mucus coating on the walls of the colon. In time this coating increases in thickness and becomes hard, like plaster on a wall. This prevents the membranes lining the colon from absorbing nutrients not absorbed earlier in their journey through the digestive system. Mucus in the colon and the toxic wastes it traps add to the level of toxins in the body as a whole, increasing feelings of fatigue and malaise. The fibre in raw food, however, gently scours away these toxic deposits. And on a diet of raw foods there is no need to eat extra fibre.

Raw foods conquer blood sugar problems

Many people who are overweight suffer from hypoglycaemia or are borderline diabetics and have difficulty in metabolising carbohydrates properly. Indeed the kind of fatigue and mental depression that comes from low blood sugar and has you drinking coffee all day or eating sweet things 'just to keep going' predisposes the development of diabetes. Raw foods can change all that. Their high fibre content has proved itself to be an important factor in normalising carbohydrate metabolism and eliminating food cravings. This is why the standard dietary approach to diabetes is rapidly shifting toward the use of more raw foods. But raw food experts such as Dr John Douglass have found that raw carbohydrates are far better tolerated than cooked ones. They don't cause the addictive craving for more that the hypoglycaemic experiences. Douglass, like the Finnish expert A. I. Virtanen, also believes that the enzymes in raw foods play an important part in the way they stimulate weight loss as they do in the treatment of obesity.

The raw slimming secret

The raw approach to slimming works with nature rather than against it. Raw slimming does not mean starvation – it means supernutrition, and slow, steady, healthy weight loss. You do not have to count calories or buy special foods or weigh out bird-size portions on kitchen scales or feel deprived when you do not eat and guilty when you do. The same raw food diet that heals cancer and diabetes and arthritis in the biological clinics of Europe does wonders for the overweight.

Some raw foods are particularly good for losing weight. Take sprouted seeds and grains (more about these in the next chapter), an ideal basis for main meals. Or fresh juices (more about these in the next chapter as well). These are an excellent way of stocking up quickly with minerals. (High protein reducing diets probably do more damage to health than any other kind because they leach so many precious minerals out of the body.) Four kinds of juice are especially recommended for slimming: carrot, spinach, beetroot and cucumber. Make freshly extracted carrot juice and add it to smaller quantities of the other three. Carrot juice is not only delicious, but also the best eliminator of toxins you could possibly find. In fact it is unrivalled in its ability to improve muscle tone and increase vitality. This is probably because it contains an abundance of important minerals plus vitamins B, C, D, E, K and beta-carotene. It also seems to restore normal responsiveness to the adrenal glands when they become overtaxed by stress or obesity.

Automatic willpower

To those who have fought the battle of the bulge many times and lost, we say 'Please do not get discouraged'. The idea of eating uncooked foods seemed an impossibility to us, until we realised that raw foods actually endowed us with the willpower to stick to them. That too may sound extraordinary, especially if you have always taken the 'Grit your teeth and think of Twiggy' approach to slimming. Living foods vibrate with a special energy which affects you physically and mentally. They give you the strength, clarity of mind, confidence and sense of wellbeing that make you want to do what is best for your body. That is something no diet of cooked foods and no amount of vitamin pills and food supplements will ever do.

10   SPOTLIGHT ON SPROUTS AND JUICES

'A vegetable which will grow in any climate, will rival meat in nutritive value, will mature in three to five days, may be planted any day of the year, will require neither soil nor sunshine, will rival tomatoes in vitamin C, will be free of waste in preparation and can be cooked with little fuel . . .' That was how Clive McCay, professor of nutrition at Cornell University, once described sprouted soya beans. They were, he declared, an almost perfect food.

Horticulture in a jar

Sprouted seeds and grains, grown in a jar in a kitchen window or airing cupboard, are the richest source of naturally occurring vitamins known. A mere tablespoon of alfalfa seeds will produce about 2 lb/1 kg of sprouts. Sprouts come in all shapes and colours, from the tiny curlicue forms of green alfalfa to the round yellow spheres of chick peas. Common seeds for sprouting are alfalfa, mung beans, aduki beans, wheat, barley, fenugreek, lentils, mustard, oats, pumpkin seeds, sesame seeds, sunflower seeds and soya beans.

Because they contain an excellent balance of amino acids, fatty acids and natural sugars, plus a high content of minerals, sprouts are capable of sustaining life on their own, provided several kinds are eaten together. They are also the cheapest form of food around. One American enthusiast calculated that he could live healthily and well on an all-sprout diet for a mere 25 cents (18p) a day. And in an age when most vegetables and fruits are grown on artificially fertilised soils and treated with hormones, DDT, fungicides, insecticides, preservatives and all manner of other chemicals the home-grown-in-a-jar sprout emerges as a pristine blessing, fresh, unpolluted, and ready to eat in a minute.

In many ways sprouts are the perfect compromise between the agriculture of years gone by and the 'just add water' mentality of the late twentieth century. All one needs is a jam jar, some fresh water and a few seeds, and in three to five days you have a marvellous mini-forest of delicious nutritional power. Such is the fantastic world of the sprout.

Far from being the new-fangled invention of food faddists and quasi-hippies, sprouts have been recognised as high quality food for almost 5000 years. They are mentioned in Chinese writings dated around 2939 BC. Szekely, co-founder of the International Biogenic Society, found references to them in documents written at the time of Jesus. Sprouted seeds are an important part of the diet of the long-lived Hunza people of the Himalayas. In the late eighteenth century Charles Curtis, a surgeon in the British Royal Navy, recorded the fact that sprouted seeds could be used to prevent scurvy, but that ungerminated seeds had no anti-scorbutic virtues. For some rather marvellous things happen to seeds when they start to germinate.

Little dynamos

A seed is a treasure chest of latent energy in the form of proteins, fats, carbohydrates, vitamins and minerals. When it is soaked in water some remarkable changes occur. Enzymes which until then have lain dormant become active; they begin to break down stored starch into simple sugars such as glucose and fructose, they split long-chain proteins into free amino acids, and they convert saturated fats into free fatty acids. The tendency that some seeds have to produce flatulence when eaten unsprouted is drastically reduced. In fact enzyme activity in plants is never so intense as at this early sprouting stage. Physicians who use freshly grown sprouts as part of healing diets claim that it is this high level of enzyme activity that stimulates the body's own enzymes into greater activity. Sprouts are, in effect, pre-digested and as such have many times the nutritional efficiency of the seeds from which they grew. Sprouts provide more nutrients ounce for ounce than any other natural food known.

Experiments show that protein levels rise with germination, and that as germination proceeds the ratio of essential to non-essential amino acids changes, providing more of those the body needs. When maize seeds germinate, for example, the concentration of lysine and tryptophan (two essential amino acids whose low levels in unsprouted corn make it a poor quality protein food if eaten on its own) increase, while the concentration of prolamine, an amino acid not necessary for human nutrition, decreases.

The vitamin content of seeds also increases phenomenally when they germinate. The vitamin B_2 content of oats, for example, rises by 1 300 per cent almost as soon as germination begins. By the time the tiny leaves form it has risen by 2 000 per cent. Other B vitamins increase dramatically too: biotin increases by 50 per cent, pantothenic acid by 200 per cent, pyridoxine by 500 per cent and folic acid by 600 per cent. The vitamin C in soya beans multiplies five times within three days of germination. In fact a mere tablespoon of soya bean sprouts contains half the recommended daily adult requirement of vitamin C. In sprouted wheat the vitamin content multiplies six times; thiamin increases by 30 per cent, B_2 by 200 per cent, niacin by 90 per cent, pantothenic acid by 80 per cent and biotin and pyridoxine by 100 per cent. Nowhere else in nature does one find such high quality nutrition at such infinitesimal cost.

Anti-cancer factors

Many years ago, when he was studying the dietary patterns of various 'primitive' cultures amongst whom cancer was virtually unknown, Weston A. Price discovered that many of their foods – millet, passion fruit and apricots, for example – were rich in a group of compounds called nitrilosides. Their cattle and sheep fed on grasses rich in nitrilosides, and so the meat and milk they provided were rich in them too.

Nitrilosides are water-soluble substances that occur in large quantities in the growing tips of seeds and young shoots and to a lesser degree in the body of mature plants. They were first isolated by Californian physician Ernst T. Krebs and his biochemist son Ernst T. Krebs Jr. The controversy over whether or not nitrilosides can be used to treat cancer still rages, with many biologically oriented physicians claiming that they cause remission and their more orthodox colleagues dismissing such claims as absurd. Nevertheless it is a fact that nitrilosides figure heavily in the diets of primitive people who suffer not at all from cancer or degenerative diseases. Dr Alec Forbes in Bristol is one of many physicians who use nitriloside-rich sprouts in anti-cancer regimes. Sprouted grains are rich in nitrilosides. Alfalfa, mung beans, aduki beans and lentils increase their nitriloside content by 50 per cent when they sprout.

Why may sprouts be effective against cancer? Ernst T. Krebs Jr explains: '. . . when they are broken down in the body they release two chemicals . . . cyanide and benzaldehyde. Body cells – the normal cells of the body – can protect themselves from such released chemicals; but cancer cells are incapable of doing this . . . both these chemicals kill unprotected cancer cells . . . Consider quickly just the nitriloside content of the diet of primitive man. He relied heavily upon the fresh succulent sprouts of the grasses, the wild legumes, millet, vetch, the lupins, wild beans and the like. Vitamin content of these plants at the sprouting stage often exceeds by twenty times or more that of the mature plant. The nitriloside content in the sprouts of some grasses and legumes is often fifty times greater than that of the mature plant. Indeed the nitrilosides and other accessory food factors that occur in prodigious quantities in the sprouting stage of the plant may be completely absent in the mature plant.'

Another major ingredient of sprouts is of course chlorophyll. This too is known to have anti-cancer properties, as Dr Chiu-Nan Lai at the University of Texas Systems Cancer Center found out when he exposed bacteria to carcinogenic chemicals in the presence of extracts taken from wheat, mung beans and lentil sprouts. Cancer development was 99 per cent inhibited.

Mineral liberation

The body can only assimilate minerals properly if they are part of organic molecules. Unfortunately the calcium, zinc and iron in peas, beans and some grains tends to be bound to phytic acid, which makes them unavailable for absorption. This is why nutritionists sometimes warn one not to eat a diet too rich in beans and seeds. Phytin is an important ingredient of many seeds – in some it accounts for up to 80 per cent of the phosphorus they contain. However, sprouting greatly lowers the phytin content of seeds, making the minerals bound to phytin available for use. At the same time it increases their content of desirable phosphorus compounds such as lecithin, which is necessary for healthy nerves and brain function. Lecithin does a lot of other useful things too. It helps to break up and transport fats and fatty acids around the body, it prevents too many acid or alkaline substances accumulating in the blood, it encourages the transport of nutrients through cell walls, and it stimulates the secretion of hormones.

And before we leave the subject of sprouts, there have been many experiments which have shown that sprouted diets have considerable rejuvenating powers. In animals they measurably affect skin and coat condition, alertness and various physiological parameters.

Raw juices for health and vitality

Fresh fruit and vegetable juices, whether drunk during short fasts or as part of a high-raw diet, have been shown to possess remarkable properties. Swedish expert on raw juice therapy Dr George Lanyi, who worked at the world famous Buchinger clinic in Germany for many years, claims that raw juice therapy can successfully treat heart and blood diseases, digestive disorders, rheumatism, diabetes, obesity, kidney and skin disorders, and problems such as anxiety and insomnia.

The tradition of raw juice healing goes back to the nineteenth century when juices were made by squeezing crushed or chopped vegetables through muslin – a tedious task. Raw food pioneer Max Bircher-Benner, for example, made raw juices the pivot of many of his dietary treatments. Max Gerson used raw juices as the cornerstone of his 'gentle' treatment of cancer, as have most other nutritionally-oriented cancer specialists. Even in the United States, where most people's idea of juice is something one pours from a tin, there is a considerable tradition in raw juice therapy. Dr Norman Walker, expert on raw foods and juices, has categorised various juices according to their nutrient content and their effect on specific ailments.

Yet orthodox medicine has always tended to dismiss raw juice therapy, refusing to accept that juices have useful healing powers. Nevertheless a chink of acceptance was discernible in a British Ministry of Health and Public Service Laboratory publication put out in the 1950s. It reads: 'Juices are valuable in relief of hypertension, cardiovascular and kidney diseases and obesity. Good results have also been obtained with large amounts, up to one litre daily, in treatment of peptic ulceration, also in treatment of chronic diarrhoea, colitis, and toxemia of gastro and intestinal origin . . . The high buffering capacities of the juices reveal that they are very valuable in the treatment of hyperchlorhydria [excessive production of hydrochloric acid in the stomach]. Milk has often been used for this purpose, but spinach juice, juices of cabbage, kale and parsley were far superior to milk for this purpose.' This interesting nugget of information is given by Dr H. E. Kirchner in his book *Live Food Juices*. Kirchner, an American physician who trained with Bircher-Benner in Switzerland, has used raw juices both on their own and with raw foods to cure such diverse ailments as failing eyesight, arthritis, infantile leukemia, anorexia and kidney failure.

However clinical studies carried out at Stanford University School of Medicine in the 1940s and 1950s proved too conclusive to ignore and the climate of orthodox opinion began to change. The Stanford researchers confirmed the fact that raw cabbage juice cures peptic ulcers, although they failed to pinpoint the vitamin or mineral responsible. The mystery factor was dubbed 'vitamin U'. As far as the healing properties of other raw juices are concerned, one can point to nearly a century of consistent clinical evidence that they work, even if few scientific studies have been done to validate the fact. Among cancer specialists there is growing awareness that fresh raw vegetables can play a preventive role in cancer. Even one of Britain's most dyed-in-the-wool establishment medical figures recently made the off-the-record comment that there is indeed some evidence that drinking carrot juice protects one against cancer.

Undaunted by the raised eyebrows of the medical profession at large, health farms and clinics throughout the Western world continue to heal and rejuvenate their clients with fresh raw juices, at a price. But the same short-term cures can be carried out at home by anyone with sufficient interest and persistence. We thoroughly recommend Kirchner's book on the subject. One of our favourite juices is carrot juice mixed with a little apple juice and the juice of a beetroot or two. We give other juice recipes and their health-giving properties on pages 277–82.

The famous Rohsäft Kur

Fasting on raw juices instead of water is probably the single most potent short-term antidote to fatigue and stress available anywhere. This is the famous Rohsäft Kur (raw juice cure) dispensed by many of the world's leading health spas.

Standard theories of nutrition have never quite been able to explain why drinking nothing but fresh raw juices for a period varying from two days to several weeks works such miracles, for 'miracle' is the word used by many of those who have had the Rohsäft experience. The visible benefits include a lessening in the number and depth of lines on the face, a firming of body contours, and healthier looking nails and hair. Certain physiological measurements change as well; if you have high blood pressure, or high cholesterol or high uric acid levels in your blood, the Rohsäft regime will bring them down.

Theories as to why raw juice fasts work tend to echo orthodox theories about aging and how to retard it. At the cellular level, you will recall, aging is brought about by a slowing down of normal metabolic processes due to an accumulation of wastes and toxins. These wastes are not simply the work of time; they can build up because of stress and/or less than ideal nutrition. Drinking nothing but freshly pressed vegetable and fruit juices is believed to counteract this build-up of wastes which gradually poisons and ages the body at a cellular level. They put the entire body through a kind of spring cleaning partly because they give one's digestive machinery a rest, partly because they speed up the body's ability to destroy dead, diseased or damaged cells, and partly because their richly concentrated cargo of nutrients helps to renew the body's flagging immune responses.

Our own experience of juice fasting has been very good indeed. We spend two or three days on raw juices whenever we feel particularly jaded, or in need of a clearer head, or if we have a lot of work to do and want to concentrate for long periods. Some weeks we juice fast for two days simply because it is pleasant to experience the balancing and energising effect that juices have on the body. We cannot recommend them too highly.

11   FOR WOMEN ONLY

Uncooked foods have enormous potential for improving the quality of a woman's life. They are one of the reasons why the world's exclusive and expensive health farms stay in business. Two weeks on a raw diet makes a woman look ten years younger – flesh is firmer, lines are softer, and skin, eyes and hair glow with health. And two years on a high-raw diet can completely transform the shape and texture and functioning of a woman's body. Even typically female problems such as stubborn cellulite, excessive menstrual flow, pre-menstrual tension and menopausal hot flushes can be eliminated on a raw diet. You do not need expensive pills, potions and plastic surgery to look and feel great. Raw foods do it better.

Nutritional deficiencies in women
Many studies carried out in Britain and the United States point to the fact that an astonishing number of women suffer from nutritional deficiencies. One three-year research project in America, referred to in connection with meso-health in Chapter 1, found that calcium and iron deficiency were widespread in women; one in two women lacked calcium, and nine out of ten were deficient in iron. And that is probably a very conservative estimate since the levels of these and other nutrients used to define health in that study were nowhere near those that a good nutritionist would recommend to anyone wanting to look and feel their best.

Many women eating the standard Western diet also suffer from zinc deficiency, particularly if they are on the pill; zinc helps to prevent stretch marks after pregnancy or weight loss and prevents skin from ugly wrinkling. Vitamin deficiencies are also common. They can be partially corrected by taking vitamin pills, but what many people do not realise is that vitamins and minerals are synergistic, each complements the function of the other. Neither vitamins nor minerals on their own will do you much good. *All* the essential nutrients are necessary if you are going to look your best. The average diet does not offer enough of these or the right balance of vitamins and minerals to build long-lasting good looks. Fresh raw foods do, particularly sprouted seeds and grains, which you can grow in your own kitchen window, and freshly pressed vegetable juices.

The special beauty nutrients

Certain vitamins and minerals which are available in far greater quantities on a diet high in raw foods than when living on a 'normal' diet are absolutely essential to maintaining shining strong hair and nails and protecting your skin from the ravages of early aging. For instance vitamin C and the bioflavinoids – those brightly coloured substances with quite exceptional properties for a woman's health and beauty – guard the health of collagen. Collagen is the fibrous protein which gives your skin its firmness and contour. Both are present in good quantities in raw fruits and vegetables. Most of these, like other water-soluble nutrients, can be completely destroyed by heat. The mineral zinc, plentiful in pumpkin seeds, sunflower seeds, and sprouted wheat, is also needed to maintain healthy collagen and for the production of new collagen. Women who have insufficient zinc (as do a large majority on the standard Western diet – particularly those taking birth control pills) tend to get stretch marks on their breasts and stomachs when they are pregnant or if they lose weight. Vitamin A, which is made in abundance in the body when you are supplied with beta-carotene from fresh green vegetables and carrots, helps regulate oil balance in the skin.

The anti-agers

Vitamin A is an important anti-ager in other ways too. Like vitamins E, C and some of the B complex vitamins it is a natural anti-oxidant which means it is important in protecting skin – indeed your whole body – from age-related changes. Anti-oxidants have ways of combatting oxidative reactions caused by radiation, chemicals and free radicals which cause damage to a cell's genetic material, proteins and lipids. It is just this kind of damage that makes skin grow old rapidly. Uncooked foods are also full of enzymes such as superoxide dismutase and catalase which help prevent the free radical damage that causes cell damage and skin aging.

What to do about cellulite

Cellulite, those ugly bumps and lumps on the thighs, bottom and upper arms which the medical profession in Britain and America dismisses as non-existent but which women find very real and difficult to get rid of, simply does not happen if you live on a 75 per cent raw diet.

The French, Italians and Germans have done a lot of research into cellulite; they know how it develops, what the special characteristics of cellulitic tissue are, and the sort of treatments that cure it. The detoxifying properties of raw foods are especially relevant to both curing and avoiding the condition. The relationship between cellulite formation and body toxicity was discovered by two French physicians, L. Meus-Blatter and G. Laroche. They found, for example, that constipation – the incomplete removal of wastes from the colon, whether or not you have the standard one bowel movement a day – was common in cellulite-prone women. So was poor lymphatic drainage – poor elimination of the wastes which lie in the spaces between cells. This inefficient removal of the waste products of metabolism and toxins absorbed from the environment is a precondition for cellulite.

Other researchers have pointed out the relationship between cellulite and poor circulation. Many women with cellulite also tend to have an underactive thyroid gland and poor liver function.

Exercise and skin brushing

Women who suffer from cellulite find it impossible to get rid of unless they make profound changes in their lifestyle, usually a radical change in diet plus lots of extra exercise.

Two things greatly speed up the disappearance of cellulite when you use them in conjunction with a high-raw diet. The first is regular aerobic exercise, and the second is skin brushing. Aerobic exercise is any exercise that gets your heart beating fast and your lungs working hard. But it needs to be practised at least three times a week for a minimum of 30 minutes at a time. Swimming, running, jogging, dancing, trampolining, cycling and digging the garden all fit the bill. Skin brushing is a well-proven European technique for speeding up the loss of toxic wastes through the skin and encouraging better lymphatic circulation. It may surprise you to know that up to one third of body wastes can be eliminated through the skin.

For skin brushing you need a long-handled natural bristle brush or a rough hemp glove. These can be found in most good chemists and health food shops. Both your body and the brush or glove should be dry. Brush the entire skin surface, except for your face, beginning with your feet, including the soles, then moving up your legs, front and back, with firm sweeping strokes. Brush from your hands up your arms and across your shoulders, then brush your back and buttocks. On your front – abdomen, chest and neck – brush a little more gently. In the abdominal area use clock-wise circular brush strokes. When your body is used to it you can increase the firmness of your brush strokes. For maximum benefit take a warm shower afterwards, then when your body is glowing with warmth switch to cold water for 30 seconds (no more). Get out and dry yourself well, and keep warm.

Stimulating lymphatic drainage

Your body's lymph system is a kind of metabolic rubbish dump. It helps it rid itself of dead cells, toxins, metabolic wastes, pathogenic bacteria, foreign substances and other assorted junk the cells cast off. Unlike the circulatory system in which the circulation of blood is controlled by the pumping of the heart, the lymphatic system has no such pump. The plasma which has seeped through capillary walls gathers in the tissues and then slowly enters the lymphatic channels – tiny vessels with one-way valves in them for carrying lymph, along with whatever small bits of foreign matter, wastes and bacteria it has gathered – through the lymph nodes and eventually back into the bloodstream.

It is the normal contractions and relaxation of muscles and the force of gravity on the body which act to pump the lymph back through its channels and eliminate these wastes. Because of its stimulating action on the tissues beneath the skin, regular skin brushing encourages efficient lymphatic drainage. It is an extraordinarily efficient technique for cleansing the lymphatic system and for clearing away waste materials from the cells all over the body as well as those – as in the case of cellulite – that have become trapped between the cells where they are held by hardened connective tissue and where they build up to create pockets of water, toxins and fat that give the skin its *peau d'orange* (orange skin) appearance. Slowly, with lots of exercise, skin brushing and raw foods, the lumpiness disappears. Exercise and skin brushing provide the pumping necessary to get the lymphatic circulation going again, and raw foods cut down the number of toxins the body has to deal with and eliminates those already stored in the tissues. Raw food also raises the level of cell vitality all over the body, eliminating the stasis that causes cellulite to form.

But raw foods do something else of even more direct benefit in cellulite: they strengthen the walls of the tiniest blood vessels in the body, the capillaries, and so reduce the quantity of blood plasma which seeps through them into the spaces between cells. It is this excess seepage which encourages cellulite, as Italian biochemist S. B. Curry at the University of Milan discovered when he examined thigh tissue from four dozen women of all ages and amplitudes and compared it with thigh tissue from women with cellulite.

Any woman with cellulite should set about strengthening her capillaries. Again it is a question of collagen, for capillaries are made of collagen. As well as making sure that at least 75 per cent of the food she eats is raw, she would be well advised to eat two or three pieces of citrus fruit (oranges, tangerines, grapefruit, lemons) a day with as much of their pith as possible because pith contains high concentrations of bioflavinoids. Capillary strength is also a factor in menstrual bleeding.

A shorter 'time of the month'

Women on an all-raw or high-raw diet often report that menstrual problems such as bloating, premenstrual tension and fatigue improve greatly after two or three months. For some of them the improvement is so dramatic that they are not aware of their periods until they arrive. This is something we discovered ourselves and at first we thought we were unique. Then we spoke to numerous other women who said they had had a similar experience. Heavy periods become lighter – a period that ordinarily lasts six or seven days can be reduced to as few as one or two. In some women, particularly those who do not eat meat, dairy products or large quantities of nuts, periods even cease altogether. What, we wondered, does this mean?

British gynaecologist C. Alan B. Clemetson, now practising in the United States, first became interested in the possibility of regulating menstrual flow with substances that occur in foods when a young Italian patient told him that she could easily cure her excessive menstrual bleeding by sucking lemons. It was the standard remedy for the problem in her home village, she said. Surprised and disbelieving, Clemetson could not quite squelch his curiosity.

A bioflavinoid link

Many years later Clemetson had an opportunity to study the relationship between citrus bioflavinoid levels in the blood and menorrhagia (very heavy periods). His research established three things. First, the capillaries in a woman's body weaken briefly just after ovulation every month and again, more markedly, for a few days before menstruation. Second, women who have heavy periods have considerably weaker capillaries than women whose flow is normal. Third, doses of citrus bioflavinoids and vitamin C over a period of three or four months significantly reduced excessive bleeding in the majority of women he tested. After his study was completed he suggested to his patients that they eat three oranges a day, with plenty of pith, because it is the pith which contains the bioflavinoids. Many of them found this was enough to maintain their lighter periods.

It seems that several of the bioflavinoids are oestrogenic, that is they mimic some of the effects of the female sex hormone, oestrogen, including oestrogen's ability to strengthen fragile capillary walls. When oestrogen levels are highest, as they are at ovulation (approximately 10 days after bleeding ceases) and again seven days later, oestrogen appears to replace the bioflavinoids in the capillary walls of the uterus. When oestrogen levels drop most markedly, as they do in the three days after ovulation and again just before and during menstruation, the bioflavinoids re-enter the capillary walls giving them some of the protection withdrawn by dropping oestrogen levels. It is because the bioflavinoids partly compensate for the fall in oestrogen that they help to reduce menstrual flow. If oestrogen levels never varied, but were always high or always low, menstruation would not occur. It is only a sustained fall in oestrogen that brings on breakdown of the uterine wall and bleeding.

Vitamin C powerfully complements the action of the bioflavinoids, but just in case you are tempted to rush to your nearest health food store for supplements instead of increasing your intake of fresh raw foods you should know that several studies show that pure ascorbic acid (vitamin C) is not as effective in treating capillary fragility and permeability as are fruit and vegetables containing the vitamin. The bioflavinoids in food greatly strengthen many of the health-protecting qualities of vitamin C. Their presence also improves the storage of vitamin C in the system.

Oestrogens in raw foods

Clemetson also studied thirty-six other commonly eaten foods in the nut, grain, fruit and vegetable category to see if any of them had oestrogen-like effects. Were any of them capable of inducing the kind of changes that occur when the hormone itself is given? He discovered that almonds, cashews, peanuts, oats, corn, wheat and apples were. When fed on these foods his experimental animals showed increases in uterine weight, increases in the volume of fluid in the uterus walls and an increase in the number of cornified cells in the vagina. Other researchers have observed similar mild hormonal effects in raw foods. When more is understood about these, special dietary treatments may become a useful alternative to oestrogen replacement therapy during and after the menopause.

Carotene and the cessation of periods

Amenorrhoea (absence of periods) in women who follow unusual dietary habits has often been attributed to high levels of carotene in the diet. Carotene is a precursor to vitamin A; it turns into the vitamin during the digestive process. Carrots, spinach and other green vegetables contain large quantities of carotene.

Medicine has long remarked that people who take in exceptionally large quantities of carotene sometimes exhibit a change in skin tone, a golden tinge rather like a gentle tan. This phenomenon, known as carotenemia, was first recorded in the British Medical Journal in 1904. It appears to have no consequences for health, apart from a general strengthening of the body's immune system. Indeed so innocuous is carotenemia that in some countries carotene tablets are sold over the counter as artificial tanners.

Recently a team of researchers from the department of obstetrics and gynaecology at Rutgers University in New Jersey studied a group of women who exhibited both carotenemia and amenorrhoea. They wanted to find out if there was a direct causal relationship between carotene intake and the cessation of periods. The normal diet of these women consisted mainly of raw vegetables, including lots of carrots. None of them ate red meat although a few ate fish and chicken. The researchers were careful to emphasise the fact that all of these women were in excellent health; amenorrhoea did not appear to affect them adversely in any way. What happened when carotene was excluded from their diet? Those women who did manage to exclude carotene-rich foods and substitute foods with little or no carotene in them resumed menstrual bleeding. Those who chose to revert back to their high-carotene diet became amenorrhoeic again.

The causal relationship between high carotene intake and amenorrhoea is fairly clear, then. But what is one to make of it? Does carotene counteract the effects of oestrogen? We know that when oestrogen levels are always low, or always high, menstruation ceases. Whatever the mechanism by which carotene exerts its amenorrhoeic effect, the lessening and eventual disappearance of menstrual flow in some women who eat a high-raw diet appears to have no adverse consequences as far as fertility and conception are concerned.

The picture that emerges, then, is that raw foods generally, and some foods in particular, namely those high in the bioflavinoids, vitamin C and carotene, reduce menstrual flow and alleviate other discomforts connected with the menstrual cycle.

An evolutionary footnote

One of the quite extraordinary claims made by women who live on an entirely raw diet – and it probably earns them the reputation of crank faster than anything else – is that menstruation is not the natural phenomenon we take it to be. Primate researchers have pointed out that the Old World monkeys do not menstruate, but that their higher relatives the baboons do; nevertheless, when fed a vegetables-only diet, female baboons cease to menstruate. Does that mean that in *Homo sapiens*, the highest primate of all, menstruation is one of the consequences of omnivorous rather than vegetarian eating habits? It is an intriguing question. If it were ever proved that menstruation is a consequence of diet many women's liberationists who regard menstruation as one of the many obstacles to women's freedom would rejoice. It would also turn our entire view of the female sexual-reproductive cycle on its head.

In the meantime the relief that a high-raw diet can offer women who suffer from any of the typically female agonies seems too important not to investigate further. As far as cellulite and general appearance are concerned the benefits of a raw diet have been well researched and are well worth making use of now.

12 THE MIND LIFT

Even more important than the way the high-raw diet made us feel physically – how it increased our stamina and energy – was the way it made us feel 'in ourselves'. In a single word: terrific.

The first thing we noticed was that those periodic feelings of discouragement which seem to come from nowhere became less frequent. Now they are extremely rare. Our thinking processes seem to be much clearer. Instead of getting caught up in emotional hassles when differences arise with other people we can stand back and see what is happening. We no longer identify so much with what we think – we feel less threatened by someone who doesn't agree. Also when one of us (Leslie) has to give a talk to a large audience she feels reasonably relaxed and confident where before she was nervous and self-conscious. This stems mainly from the fact that she gathers her thoughts more easily than she did before. The other partner (Susannah) used to suffer from minor bouts of depression for no very obvious reason. After several weeks on a high-raw diet these stopped coming, but it only took a weekend of old-style eating to bring back one of her familiar lows.

A biochemical key to even temper

We both find that raw food keeps us much more even tempered. It gives us a sense of physical–psychic balance we did not have before. Life on a high-raw diet is not the endless seesaw of minor ups and downs we once thought it. This makes us wonder if many of the negative feelings we all get from time to time are not so much psychological in origin as physiological, a sign that body chemistry is out of balance and toxins are building up. In our experience, the longer you stay on a high-raw diet, the more positive you feel about yourself and life in general.

That is not to say that we now live in some blissful altered state of consciousness, like beings from a rosier world. Far from it. We find ourselves more *involved*. We have more energy and we find more enjoyment in the things around us, from a new flower in the window box to a novel or a piece of music.

Fulfilling the blueprint

Slowly we realised that this experiment in high-raw eating which we had rather naively begun was turning out to be more all-encompassing in its effects than we had ever imagined. We came across what Dr Ralph Bircher had once written of Bircher-Benner, who believed that raw foods could not only help cure his patients of illness but also formed an important part of a self-realising and self-healing system which could help his patients in every conceivable way to fulfil their individual potentials. And we began to understand how this might be so. He said: '. . . he initiated a school of medical thinking, treatment and outlook which envisaged, and still envisages, the patient as an indivisible whole, as a psycho-physical personality, in order to promote the realization, as far as possible, of the potentialities and original "blueprint" given him at his creation . . .' A high-raw diet, he insisted, helped people fulfil their potential in every area of their lives.

But again that same question kept arising: why? Why should such a regime make us feel so different? Why should we now find ourselves better able to cope with stress, feel less fatigue, and even (we discovered after a few trips across the Atlantic) be able to avoid the lion's share of the damnable jet lag which used to leave us feeling wiped out after crossing several time zones?

Raw foods counter fatigue

Chronic fatigue, marked by irritability, lethargy and the feeling that almost everything is too much bother, has been called 'the plague of modern civilisation'. A certain wit once observed that 99 per cent of the world's work is done by people who feel under the weather. However only 20 per cent of people who go to their doctor complaining of tiredness are diagnosed as suffering from something detectable. Anaemia is one of the commonest detectable causes of tiredness. Other causes may be under-production of certain hormones (as in diabetes, hypothyroidism and hypopituitarism), chronic infections, and occasionally heart disease, especially if there is valve damage which prevents the heart from pumping enough oxygenated blood around the body. But the other 80 per cent of those who feel chronically tired are told there is nothing wrong with them according to laboratory tests and X-rays. They go home as tired as they came.

A deficit of the minerals potassium and magnesium is known to be a major cause of fatigue. Our high-raw diet, because it is rich in fresh green vegetables and sprouts, is also rich in chlorophyll, therefore in magnesium, for chlorophyll is built around a core of magnesium. It is also, in contrast to the average Western diet, very high in potassium.

Any long-term physical effort requires large reserves of potassium.

In the 1960s an American physician, Dr Palma Formica, was curious to know what effect magnesium and potassium supplements would have on fatigue. Her volunteers for the experiment were a hundred chronically tired people, 84 women and 16 men. For five or six weeks she gave them extra potassium and magnesium. Then she recorded her findings: 'The change was startling. They [became] alert, cheerful, animated and energetic, and walked with a lively step. They stated that sleep refreshed them as it had not done for months. Some said they could get along on six hours a night, whereas formerly they had not felt rested on twelve or more. Morning exhaustion had completely subsided ... Several of the husbands called and expressed appreciation for the physical improvement and consequent increase in emotional wellbeing of their wives.' Eighty-seven of Dr Formica's hundred volunteers improved, even though some had suffered from debilitating fatigue for two years or more.

Another reason why a high-raw diet so successfully counters fatigue and lifts the spirits is that it affects blood sugar levels. It is now widely accepted that functional hypoglycaemia (low levels of sugar in the blood), which is believed to be widespread, is responsible for that mid- to late-afternoon tiredness peak that affects so many people and has them reaching for strong cups of coffee and sweet snacks 'just to keep going'. The same low blood sugar levels are probably responsible for many feelings of lowness and depression too. Dr John Douglass and others who recommend a high-raw diet for diabetics do so because raw fibre helps to stabilise blood sugar levels. It does the same for someone suffering from persistently low blood sugar levels, and abolishes the mood swings and other symptoms that characterise hypoglycaemia.

Help for allergies and addictions
Another cause of mood swings can be food allergies. It is not uncommon, for example, for wheat and milk products to cause catarrh, digestive problems and feelings of chronic lethargy and tiredness. A lot has been written in the popular and scientific press about food allergies and special diets to deal with them, regimes in which the 'offending' food is left out. What few people are aware of (although Bircher-Benner discovered the fact more than 50 years ago) is that a tendency towards food allergies is significantly reduced by a high-raw diet based on vegetables alone. This means a vegan-type diet in which flesh foods and dairy products such as milk and eggs are banned. Other allergies such as skin rashes, hay fever and rheumatic conditions can also be reduced by a vegetables-only regime.

Allergists such as H. Rinkel and T. Randolph and others have written extensively about the addictive aspects of food allergies: the allergic person actually develops a craving for the very thing he or she is allergic to. This craving masks the allergy provided the body's resistance is high enough, but when resistance weakens all the symptoms of the allergy develop. Douglass found that a high-raw diet is a very effective weapon against allergies and the addictive patterns that accompany them. Even common addictions such as cigarettes and alcohol seem to lose their force after a few weeks on a raw diet. At first Douglass was most unwilling to believe that raw foods lessen addictive cravings, but his patients insisted that after a few weeks of high-raw eating they simply did not want as many cigarettes or drinks as before. Willpower did not come into it. Douglass concluded that, in some mysterious way, raw foods must sensitise the body both to what is good for it and what is bad for it. Experimenting with specific raw foods and their effects he found that some, such as sunflower seeds, were particularly effective in depressing the cravings associated with addiction.

Sunflower seeds versus nicotine

Sunflower seeds are an excellent source of vital nutrients. They contain most of the B vitamins, vitamin E and also many essential fatty acids, and pound for pound twice as much iron and twenty-five times as much thiamin (a B vitamin) as steak. Douglass found they were particularly good for people trying to wean themselves off cigarettes, so much so that he now recommends would-be non-smokers to carry a handful of raw shelled sunflower seeds around with them and every time they feel the desire to smoke to pop a few into their mouth and munch until the desire subsides. In a few weeks, he says, even the desire to smoke seems to fade. How does he explain this David and Goliath effect, the humble sunflower seed versus the ogre nicotine?

It seems that sunflower seeds contain ingredients which mimic some of the effects of nicotine. To an extent therefore they give smokers some of the gratification they seek from nicotine. Nicotine tends to have a mildly soothing, sedative effect on the nervous system; so do sunflower seeds because they contain various sedative essential oils, and also plenty of B vitamins, always good for the nerves. Nicotine triggers the release of glycogen from the liver, producing a temporary increase in brain activity; sunflower seeds produce a similar lift. Nicotine raises the level of adrenal hormones in the body as well; sunflower seeds also stimulate the adrenal glands. Sunflower seeds are non-allergic, and effectively break through the smoker's pattern of addiction without themselves becoming the target of a new allergy.

In times of stress

Another thing we particularly enjoy about our high-raw way of living is that it has made us much more resistant to stress. Our resilience is not destroyed by having to drive in rush hour traffic or stay up all night to finish a piece of work. We seem able to gear up or gear down to whatever is demanded of us. When we asked various nutritional experts why this was, they said it was probably due to the way in which raw foods affect the body's acid-alkaline balance.

Balanced body chemistry is not merely a recipe for keeping calm and collected, but a fundamental necessity for health. Over-acidity lies at the root of many illnesses, particularly arthritis and rheumatism. Every food you eat tends to be either acid-forming or alkaline-forming. If your diet contains a lot of sugar, coffee, meat and other concentrated proteins, processed foods made from white flour, and only a few fresh vegetables and fruits, you are consuming foods which are mainly acid-producing and you will tend to feel stressed very easily. We used to feel pretty edgy ourselves whenever we reverted to 'a good English breakfast' of eggs, bacon, toast and coffee. You get this nervy feeling because your body has used up its alkali reserves in an effort to balance the acid-forming foods you have eaten.

Unfortunately the compounds that the body produces in response to stress are also acidic. A combination of acid-forming foods and periods of stress sends the body's acid levels up and up. So it is important, for overall health and as an antidote to stress, to eat plenty of alkaline-forming foods. A desirable ratio of alkaline- to acid-forming foods would be 80 per cent to 20 per cent, four to one in favour of the alkaline. When you know you are going to be exposed to situations you find stressful, eat an even greater percentage of alkaline-forming foods.

A cure for jet lag

To our mind one of the most unpleasant and unavoidable forms of stress is jet lag, when the body's inner biochemical rhythms get out of synch after crossing several time zones. This does not happen when you fly north or south, of course, only if you fly west or east. The symptoms of jet lag include confusion, exhaustion during the day and sleeplessness at night, and that awful 'spaced out' feeling of not quite knowing where you are. When we began eating a high-raw diet we noticed that we felt all these unpleasant things to a much lesser degree. Then we began to experiment with various ways of eating before, during and after flights to see if we could improve things further. Finally we hit on a method which works wonders for both of us. Others who have tried it report similar relief from time zone troubles.

This is what we do. The day before a flight we eat lightly, and raw foods only, with the emphasis on salads and fruits, which are alkaline-forming, rather than nuts and seeds, which are more acid-forming. On the day of the flight we eat nothing at all. Instead we drink lots of water or take some fresh fruit or vegetable juice on the plane with us. This does two things: it helps your digestive system prepare itself for the change in mealtimes at the end of your journey, and keeps you from having to eat the pretty horrendous meals most airlines serve. The day after the flight we eat raw food only. The following day we go back to our normal way of eating, which is about 75 per cent raw. We end up able to sleep at night, well oriented in time and space, and able to work productively – in short feeling like human beings rather than wrung-out dishcloths.

Is raw energy a key to self-fulfilment?

Five years ago such an idea would have made us both laugh out loud. We would have dismissed the notion as nothing more than the fantastic invention of some strange Californian cult. But when we began experimenting with raw foods and found they affected us so profoundly we started to wonder. We ploughed through a number of ancient treatises on the relationship between diet and the mind from the Vedic teachings that form the basis of India's traditional Ayurvedic medicine to Szekely's translations of Essene teachings. We were not altogether surprised to find that the kind of diet most of these writings recommend for increasing mental and spiritual awareness is one very close to what we were eating.

If we had not read so many scientific reports of meticulously conducted research we might have been tempted, as some of our friends are, to dismiss the positive effects of raw foods as yet another demonstration of the placebo effect – 'a little of what you fancy does you good'. But the clinical and experimental evidence, mountains of it, suggest otherwise.

The effects of raw food on the body are a matter of biochemistry, not of the imagination, unless of course rats, mice and guinea pigs are influenced by placebos. No one has ever dared suggest as much.

So now we are not so sure. We certainly feel that the high-raw diet has made it possible for us to work harder and more effectively than ever before, that it has given us a more equable attitude to other people and heightened our senses so that we get a lot more enjoyment from things around us. Bircher-Benner's statement about how such a diet was an important tool for self-realisation and self-healing which 'helped the patient in every conceivable way – not just by banishing the symptoms of disease' no longer sounds as far-fetched as it once did. So we keep an open mind. Raw foods may indeed help human beings to fulfil their potentials, not only for high-level health and good looks but in many other ways too. We figure only time will tell.

THREE

THE
RAW ENERGY
WAY OF LIFE

13 SWITCHING TO A HIGH-RAW DIET

'A high-raw diet? I could *never* stick it!' Admittedly it does sound pretty extraordinary. But what most people don't realise is that by eating raw food you gradually come to feel that the high-raw diet is not just what's *good* for you, but what you *like* best. Living foods vibrate with a special quality of energy. This energy, when regularly taken into your body, changes you physically as well as mentally, bringing you strength, clarity of mind, confidence and a sense of well-being which make you want to do what is best for your body. It also heightens your senses so that the smells and tastes and textures of foods become a source of growing delight. Before long a large piece of pizza and a rich chocolate dessert lose their appeal.

The way one feels eating mostly living foods compared to mostly cooked, convenience and junk foods might be expressed by the following analogy. When you go out to a smoke-filled night club or disco you may have energy and feel lively, but it is a sort of nervous frenetic hyperactivity; that's what happens when you eat a high protein diet or junk food. When you are out walking by the sea or in the mountains you feel a different kind of energy – exhilaration – where all your body cells seem to radiate life; this is the effect raw food has on your body. The high-raw way of eating is not just a diet, it's a way of life. Most of us have at least one of the following goals:

to be happy
to be healthy
to look beautiful
to be slim
to be fit
to stay young
to be successful.

Extraordinary as it may seem, a high-raw diet can help

you achieve these goals. Often we are held back from fulfilling our potential by feelings of negativity about ourselves; we think that happiness and success are somehow someone else's birthright and that we are doomed to misery and despair. The detoxification potential and the high nutritional potency of the high-raw diet can actually alter the body's biochemistry, improving not only your physical state, but even your outlook on life. It can help dispel these feelings of hopelessness as well as giving you more physical and mental energy to fulfil your new ambitions. A high-raw diet is not just for the sick or unhappy – and it would be a shame to wait for sickness or depression to befall you before discovering it.

Of course there are those who do not want to be healthy, who would rather drink, smoke, eat rubbish and die young. No doubt they would prefer a book called *Sex, Money and Success: how to have more by doing less*, first edition complete with tiny miracle pill! These people are probably most in need of Raw Energy; little do they realise that their hunger for stimulation and sensation is a result of the abuse and misuse to which they subject their bodies. They desperately need the inner vitality that a high-raw diet brings.

The 75 per cent rule

It is unhealthy to become fanatical about a way of eating to the extent that it runs your life. The Raw Energy way of life is not a deprivation where you forbid yourself anything and everything cooked – for social reasons alone that would be very difficult to sustain. When we recommend a 75 per cent raw diet we are not compromising but being realistic. We have found that a diet of about 75 per cent raw food plus some wholesome cooked foods works best. If sensibly prepared some cooked foods have considerable nutritional value, and they also add variety to your diet. But it is important to stress that the focus of your main meals should be the raw rather than the cooked element, and most emphatically each meal should begin with raw food.

Cook but don't cremate!

On pages 285–6 we list some of the cooked foods we occasionally include in our mainly raw regime. Heating any food above 130°F (boiling point is 212°F) inevitably destroys its enzymes, but with careful cooking most of the minerals and some of the vitamins can be saved.

The worst thing you can possibly do to vegetables is boil them for half an hour until they are mushy, strain the water down the sink, and serve. Who wants mere skeletons with virtually no taste or goodness in them? The best method of cooking vegetables is steaming. There is no water to leach out minerals and vitamins and they retain most of their colour, shape and texture. You can buy a vegetable steamer complete with saucepan and fitted basket or just a perforated metal basket which fits all sizes of saucepan. Or you can improvise your own steamer with a metal collander or sieve and a saucepan lid. Quarter fill the saucepan with water, bring it to the boil, put the vegetables (chopped or whole) into the basket and cook with the lid on. Your vegetables will cook in just a few minutes because steam is hotter than boiling water. If you must boil food, rice for example, only use just enough water.

Stir-fried vegetables also cook quickly without losing too much of their goodness. They should be eaten crisp, not soft. Pre-heat a very little olive oil in a wok or heavy frying pan, add your chopped vegetables or sprouts, and cook for just three or four minutes, stirring with a wooden spoon. A touch of soy sauce goes very well with stir-fried foods. When frying other foods use the minimum of oil, preferably olive oil, never heat it to smoking point, and remember that small pieces require less frying and therefore lose less goodness than big ones.

All meat, game, fish and poultry should be cooked as slowly as possible so that they retain their juices. This also conserves vitamins and minerals. Never sprinkle anything with salt before you cook it as this very effectively draws the juices and minerals out of it!

Easy does it . . .

It is not necessary to throw yourself madly into an all raw diet in order to reap the benefits of raw energy. A 75 per cent raw diet will suffice, and even that is a way of living that needs to be entered into with respect and sense. Few people who have been living on an ordinary Western diet can suddenly change to a mainly raw diet without some unpleasant effects. If you give your body *carte blanche* to throw off all the toxins and wastes that have accumulated over the years they flood into the bloodstream causing what nutritionally-oriented doctors call a 'dynamic healing crisis', which can result in headaches, aches and pains, tiredness and irritability. Depending on the state of your body these cleansing reactions may be severe, mild or go completely unnoticed.

Raw foods should be introduced into the diet slowly. Start by replacing one of your normal meals each day with a large fresh raw salad and experiment with drinking vegetable or fruit juices or herb teas instead of coffee, tea, alcohol and soft drinks. Gradually eliminate suspect foods (more about these on page 166) and cut down on stodgy cooked foods (bread, cakes, pasta) until you reach a balance of about 75 per cent raw and 25 per cent cooked. This is the best way to discover the benefits of raw energy.

If, however, you are very keen to change to a high-raw diet as fast as possible, or if you would simply like to experience its spring-cleaning effects, then try our Ten Day Prove-It-Yourself Raw Energy Diet. But remember it is intended for the basically healthy. If you have the slightest doubt about your health, consult your doctor.

Ten Day Prove-It-Yourself Raw Energy Diet

Diets always begin tomorrow. And tomorrow they begin the day after. It is hard to find a diet that fits your daily routine, and so the good intention gets put off and put off, sometimes forever. This ten-day Raw Energy diet is different. It is designed to begin on a Friday evening, spread over a weekend, a week and another weekend. Even if you have a nine to five job, or are obliged to eat at least one of your main meals in a restaurant, it can be followed with ease.

The Prove-It-Yourself diet is an introduction to the pleasures of raw eating, and also a quick and effective way to cleanse your system and lose a little weight – uncooked foods have a way of making excess weight fall away. Almost certainly it will give you a taste of the state of wellbeing that we enjoy on a mainly raw diet.

Skin brushing, already mentioned on page 113, is a technique which complements the cleansing effects of the Ten Day diet. While the diet ensures elimination of toxins internally, skin brushing ensures that as many toxins as possible are eliminated through the skin. As an added bonus it increases lymphatic drainage and improves muscle tone, especially if you take a warm then a cold shower afterwards.

Day 1 *(Friday)* Pre-diet day

The purpose of this pre-diet day is to prepare your body for the coming change of diet. On this first day no stimulants (coffee, tea) or depressants (alcohol) should be taken. Bread and cooked carbohydrates (pasta, cereals) should also be avoided. Make your last meal of the day a large raw salad of vegetables and fruits. This is also a good time to try growing your first sprouts so they will be ready from day 4 onwards (full instructions on page 183) if you cannot find them in your local shops.

Days 2 and 3 *(Saturday and Sunday)* Fruit fast

The fruit fast is one of the best ways of clearing your system quickly. Because the effect is so dramatic you may find that you experience some mild elimination reactions such as headache, irritability or tiredness at some point within the first three days. For this reason Days 2 and 3 coincide with a weekend so that you can rest as you feel necessary.

The fruit fast is effective in several ways. In a purely physical sense fruit is mildly laxative and a wonderful intestinal 'broom' to sweep your alimentary canal clean. Fruit is alkaline-forming; most stored wastes which are responsible for aches and disease in general are acidic. When your body is given the chance to throw off these wastes, as it is on the Prove-It-Yourself diet, they first enter the bloodstream. The alkalinity of the fruit helps to neutralise them so that they are not harmful and can be quickly expelled. In this way you minimise the possibility of any cleansing reactions and rapidly achieve a better acid/alkaline balance. Fruit also has a high potassium content. This is helpful in ridding the system and the tissues of excess water, increasing oxygenation in the cells and raising microelectrical potentials.

Many people experience no cleansing reactions at all with fruit fasting, but if you do there is no need to worry. They are a perfectly normal consequence of the rapid mobilisation and release of stored toxins and wastes. If you get headaches, or muscle or joint pains, or feel more than usually sensitive, tired or emotionally unsettled, it is not because two days of fruit fasting is harming you but because it is undoing harm stored up for many months or years. Get plenty of rest and plenty of fresh air. Deep breathing also helps your body to get rid of wastes. Though jogging or running is normally a good way of mobilising and eliminating toxins, strenuous exercise during Days 2 and 3 is not advised. Your body is working hard enough cleansing and renewing itself. And remember that Raw Energy works not only on the body but on the mind and spirit too. You are dieting for the sake of the whole you, not merely for the sake of your body.

For Days 2 and 3 you choose one kind of fruit and eat it throughout the day. Each fruit has special health-giving properties. We find that apples, grapes, pineapple, pawpaw, mango and watermelon are particularly good when we fruit fast. Here are some notes to help you make your choice.

Apples Excellent for detoxification. They contain galacturonic acid which helps to remove impurities from the system, and pectin which helps to prevent protein putrefying in the intestines. Because apples contain a lot of fibre they are great 'brooms'. They are also good for strengthening the liver and stimulating digestive secretions and are rich in vitamins and minerals.

Grapes Very effective cleansers for the skin, liver, intestines and kidneys because they discourage the formation of mucus in the gut. Grape sugars are a quick source of energy because they are easy to assimilate. Grapes are also good blood and cell builders.

Pineapple A concentrated source of bromelain, an enzyme which activates the hydrochloric acid in your stomach and helps to break down protein. Pineapple is also believed to soothe internal inflammation, accelerate tissue repair, stimulate hormone production and clear away mucus in the gut.

Pawpaw and mango Rather expensive and sometimes difficult to find but rich in an enzyme called papain (mango less than pawpaw). Papain resembles the enzyme pepsin in the stomach and, like bromelain, helps to break down excess protein. Both fruits are good for cleansing the intestines and helping digestive disorders. Mangoes are also believed to relieve depression.

Watermelon A wonderful diuretic (increases urine excretion) and great for washing your system clean. It is used to ease stomach ulcers, lower high blood pressure and soothe the intestinal tract. To get the benefit of the chlorophyll-rich skin and vitamin-packed seeds juice them with a little of the pink flesh and drink about half an hour before you eat a melon meal.

Only one kind of fruit is eaten throughout the day because this is least taxing for the digestive system (it is also, incidentally, the best way to lose weight). However if you are limited as to the amount of a certain fruit available you can change fruit half way through the day provided you leave a gap of at least two hours before starting the new one. How much fruit you eat is up to you. Eating fruit by itself will not make you gain weight. You will find that you need to eat more frequently than usual as fruit is digested very quickly and does not remain in the stomach for more than an hour. We suggest eating no more than four or five fruit meals spread throughout the day (eating continually is tiring for the digestive system).

Days 4, 5, 6, 7 and 8 *(Monday to Friday)* **Replenishing**
During these five days the cleansing process continues but your newly cleansed cells are given all the nutrients they need to fortify and rebalance your system. The vitamins, minerals and enzymes in the raw vegetables and sprouts you eat prod sluggish cells into action and boost the efficiency of all your body systems. Each day begins with a fruit breakfast, very important because it encourages the liver (most active early in the day) to continue rapid elimination of stored wastes. Lunch is a large raw salad with sprouts, and a topping of seeds or blanched almonds. Dinner is steamed or stir-fried vegetables with a blended raw topping. Swop lunch for dinner and dinner for lunch if it is more convenient. What about taking a bag of ready prepared fresh vegetables or sprouts to the office, along with a jar of dressing to dip them into? If you eat lunch in a restaurant ask them to prepare you a mixed raw salad with a sprinkling of olive oil and a squeeze of lemon, or just order steamed vegetables, without butter. Difficult sometimes, but not impossible.

Days 9 and 10 *(Saturday and Sunday)* **Reorientation**
The last two days help you to adjust to a diet containing more cooked food, and set you on the road to a 75 per cent raw way of eating. The focus of each meal is still raw, but one meal a day (lunch or dinner, whichever is convenient) includes a cooked piece of game, poultry or fish or a vegetarian casserole.

Ten Day Prove-It-Yourself Raw Energy Diet
(see Recipe Section for inspiration!)

Day 1
Dinner
Fresh raw salad only, eaten early in the evening to give your system a good 12 hours to start eliminating before your first

meal on Day 2. No tea, coffee or alcohol. Cup of herb tea before going to bed.

Days 2 and 3
First thing
Mixed fresh juice of one orange and half a lemon, topped up with spring water, *or* a cup of herb tea with a squeeze of lemon (lemon verbena is a very good 'wake-up' tea – sweeten with lemon if you like).

At three or four hour intervals
A fruit meal. Choose from apples, grapes, pineapple, paw-paw, mango or watermelon, one fruit for Day 2 and another for Day 3. Make things interesting for yourself. Grate, slice or dice your fruit; or chill it and put it through the blender, perhaps adding a little water, and serve with crushed ice and a dusting of cinnamon, allspice, nutmeg or ginger. Or juice your fruit for one or two of your meals.

Drinks between or after meals
Herb tea or spring water only.

Days 4, 5, 6, 7 and 8
First thing
Fresh orange and lemon juice *or* herb tea, as for Day 1.

Breakfast
Fruit (not bananas). One kind of fruit only *or* a simple fruit salad dressed with a little lemon juice, honey and spices, *or* a delicious blender fruit shake, such as a 'mango smoothie' (mango flesh plus freshly squeezed orange juice).

Lunch
A large raw vegetable salad. Invent new combinations each day. Top with generous helpings of sprouted seeds, pulses and grains, and dress with olive oil and lemon, olive oil and cider vinegar, or avocado or mayonnaise dressing (see recipes), and lots of fresh herbs such as basil and parsley. An

avocado is an excellent addition to such a salad; so is a sprinkling of sunflower, pumpkin or sesame seeds, or a few blanched almonds.

Dinner
Steamed or stir-fried vegetables seasoned with fresh or dried herbs and perhaps a little soy sauce. Use three or four different vegetables together. Add a few sunflower, pumpkin or sesame seeds, or almonds or pine nuts, and spike with spring onions. Make a delicious sauce such as Curried Avocado Dip to pour over the top.

Drinks between or after main meals
Fresh fruit or vegetable juice or herb tea sweetened with honey.

Days 9 and 10
First thing
As for Days 1 to 8.

Breakfast
As for Days 4 to 8.

Lunch
As for Days 4 to 8.

Dinner
Fresh vegetable juice *or* slices of raw vegetables plus a dip. Then 4oz (130g) poached or grilled fish, *or* the same quantity of lightly fried lamb's liver *or* chicken slow-roasted without the skin *or* vegetarian casserole (beans, peas, lentils and vegetables). Steamed vegetables, and boiled brown rice if desired. A green salad. A piece of fresh fruit for dessert (optional).

Drinks between or after main meals
As for Days 4 to 8.

How to introduce children to raw eating

The 'carrots are good for you' approach is usually a dismal failure with children, especially younger ones. After all, adult incentives such as rejuvenation, easing aches and pains and being razor-sharp for business meetings cut no ice with six-year-olds! But switching to a mainly raw diet is bound to affect the family and there are ways, as we have discovered, of making the switch a fairly painless one.

Wouldn't it be wonderful to have a child who didn't like sweets or junk food and only ate the foods that were really good for him without having to be told 'Now Rupert, if you don't eat your cabbage and spinach you'll never grow up to be big and strong like your father.' Actually all breast-fed babies have a natural affinity for wholesome foods from birth. It is only when their taste becomes adulterated by commercial baby foods loaded with sugar and salt that their natural discrimination falters and they develop a craving for the wrong kind of food, usually sweet things. It is interesting to observe that young children instinctively seem to know which foods they need. They will often go for one type of food for several days because it contains the proteins or nutrients that they need at that stage of development. So if you can start your child on raw energy foods from the very beginning, so much the better. Aaron, the three-year-old member of our family, likes nothing better than to munch his way through a red or green pepper, drink a glass of freshly made carrot and beetroot juice, or chew on a piece of fresh seaweed. And he is not unusual.

The problem arises when you try to convert older children to a mainly raw diet. Their tastes have already been partly formed. Almost any attempt at conversion is likely to meet with opposition at first. It certainly did in our family several years ago when we had a minor war of children versus parents which resulted in a compromise to eat 'Mommy's kind of food' twice a week and 'normal' food the rest of the time. In fact Mommy's food very soon took over and after a few tearful episodes over half eaten bowls of salad, with the threat of salad for breakfast, the children soon learned to eat and even enjoy a totally new diet.

The important thing to remember is that all likes and dislikes are learnt. So new likes and dislikes can also be learned. As an extreme example, we once met a spoilt child who only liked artificial fruit flavouring and nearly threw up when we told him to eat his fresh strawberry muesli! With some children it is enough to tell them how healthy raw foods are, or how damaging junk foods are. Branton, the eldest 'child' in our family used to be an avid sweet eater until the dentist told him that if he continued to eat sweets like there was no tomorrow he would have false teeth by the time he was twenty. He has hardly eaten a sweet since! With other children one has to resort to sneakier tactics.

Acting underhand

Getting children to act in their own best interests takes patience and persistence. It is important to set a good example yourself and introduce raw foods gradually. If children see you enjoying more salads and fresh fruits they will eventually come round, particularly if you avoid making too much of an issue of it. Here are a few 'encouragers'.

Snacks   These are often the worst offenders in children's diets, but they need not be. We remember seeing an American television commercial in which a child scrubbed and cut carrots into sticks and refrigerated them in a jar of cold water to munch on. Marvellous, we said! Crudités, raw sliced or whole vegetables of all kinds, make wonderful snacks (see Crudités on pages 211–12). Children love the idea of crunching on fresh vegetables. When the children in our family were young we used to pull up carrots and radishes which Dumpa (grandfather) grew in his garden, hose them off, and eat them ... nothing ever tasted as delicious! Crudités are also good as part of formal meals because children love the chance to eat with their fingers; that way they get the feel as well as the taste of their food.

It is a good idea to keep a bowl of fresh fruit and a bowl of nuts (in their shells because they are fun to crack!) in the kitchen for children to help themselves to between meals instead of going to the shop for a bag of sweets or a bar of chocolate. You can make delicious wholesome snacks of frozen bananas dipped in carob powder, chopped nuts and honey, and served on sticks, or ice lollies made from frozen orange juice concentrate and plain yoghurt.

Packed lunches A raw packed lunch can consist of a wide range of delicious and nutritious items such as: Essene bread sandwiches or 'crisps'; cheese; seed or nut cheese; raw patties; whole or chopped vegetables; sprouts; pieces of fruit; nuts and seeds; fresh juice in a thermos (see Recipe Section for more ideas). Sprouts are one of the best foods to introduce children to because they are such fun to grow and eat (see page 183 for sprouting hints). Why not put the children in charge of sprouting and let them take pride in growing all the different sprouts you want to use? They will be fascinated to see how the baby seeds develop – few things are as exciting as the miracle of germination.

Goodies These are a very satisfactory replacement for sweets as they are undeniably delicious (see pages 267–70 for sweet treats).

The important thing is to convert your children gradually without them feeling that they are being forced into anything. Rather, think of them as being on your side trying out a delicious and interesting new way of eating. If you meet resistance use a little gentle persuasion and remember when they are old enough to have children of their own they will be grateful for your efforts and persistence.

14 EATING AWAY FROM HOME

It is tempting simply to abandon raw food as soon as you have to eat away from home. It's too much hassle and you don't want to 'put anyone out'. So you take the path of least resistance and eat whatever happens to be going. Having followed this path ourselves for many years and done our best to oblige friends and hosts and chefs in restaurants, we have come to the conclusion that in terms of how you end up feeling after an unwanted meal, it simply isn't worth it. Far better to eat only what your body really wants and to learn a few rules of the game to get you through sticky situations without causing offence, commotion or embarrassment.

In restaurants

Restaurants are the easiest away-from-home eating situations because you can choose what you want. Nevertheless you have to know what you are likely to get or you can be in for a shock. The better the restaurant the more obliging they will be. Most vegetarian, Italian and French restaurants, and restaurants with salad bars, are pretty reasonable for fresh foods. If you have the choice, steer clear of Chinese restaurants – the sugar, oil, salt and monosodium glutamate in the dishes can wipe you out. Indian restaurants also tend not to have many fresh vegetables.

● Be very wary of sauces, dressings, vegetables swimming in butter, and fried anything – often it is the oil, cream and white flour which accompany overdressed restaurant food that make you feel so lethargic afterwards.

● Ask for salads undressed and a little oil and fresh lemon to make your own dressing. This way you can add as much or as little as you want. Often you can ask for a salad to be made up specially for you.

● Try to order the same number of dishes as the other people you are eating with. Begin with half an avocado or a bowl of soup (clear soups are your best bet), order a salad as your main dish (you can always order a starter as your main course if the entrées are unappealing), a piece of fresh fruit for dessert, and a herb tea or fruit juice instead of coffee.

● Don't be afraid to drink a good glass of wine with your meal if you want one, but not an alcoholic cocktail as well – it will only impede your digestion. Have a fruit juice as an aperitif instead.

● Eat slowly so that you don't finish earlier than your

companions. Seeing you watching them over your empty plate is bound to make them uncomfortable. If you order the same number of dishes as they do, even if you have a starter as your main course, your 'unusual' menu will probably not be commented on.

● Be prepared to put up with a little teasing. Most people feel uncomfortable sitting at the same table as someone who appears to be dieting or looking after their body while they recklessly abandon health in favour of 'fun'. This discomfort can show itself in disparaging remarks about 'rabbit food', or as simple curiosity. But it does get boring to be continually asked 'Why do you eat that kind of food?' This question can be parried in various ways. You can say 'for medical reasons', which in polite company is enough to change the topic of conversation, or simply 'because I like it'. To dispel the idea that you are using fanatical will power to stay on a diet just because it is healthy a useful answer is 'Well, I *can* eat any kind of food, but I just find I feel so much better on mostly vegetables and fruit.' It is best to give a short answer and then change the subject. If the questioner is sincerely interested in why you eat raw food you can go into greater detail. Avoid getting into arguments with wise guys and trying to justify your preferences. It simply isn't worth it. You are not obliged to justify anything. Just keep quiet, smile politely and continue munching.

At dinner parties

Going to dinner with friends is a little trickier because you are not handed a menu to choose from. Generally you know beforehand what type of food to expect. If you know you are going to be regaled with a rich heavy meal and very little fresh food it is a good idea to take the edge off your appetite before you leave home with some crunchy crudités. That will make it a lot easier to plead lack of appetite or fake eating along with the rest of them. Here are some more helpful tips.

● Eat plenty of the side salad (?) and steamed (?) vegetables. Go easy on the main course if it is rich but a little won't harm you. Avoid dessert altogether, especially if it is a heavy pastry and cream concoction. Say the main course was so delicious that you're full or ask for a piece of fresh fruit instead.

● Don't feel bad about leaving what you don't want. It is better to push those greasy French fries around your plate than force them down. ·

● If your host is a good friend, ask if you can make a dish to bring along, a salad for example.

● You will end up eating less than everyone else, which is a blessing if the food is very rich. No one will really notice as long as you eat slowly. Remember that chewing your food really well not only gets the maximum goodness out of it but also makes you feel fuller on less.

● Don't be afraid to say 'no thank you'. You can always add 'It looks lovely, but I'm afraid I'm rather full.'

● The better you know your host, the easier it should be. Try to avoid potentially difficult situations by suggesting 'drinks' or 'tea' rather than a meal. No one minds you turning down their whisky, but their lovingly prepared pasta or trifle . . .?

● Above all be consistent. Take yourself seriously and put your needs first. If you are serious about what you are doing people will make fewer jokes at your expense. Respect yourself and others will respect you too.

On journeys

Travelling anywhere, especially long plane journeys, is likely to upset your normal routine. Often just the excitement of going somewhere on holiday or the pressure of an important business trip is enough to derail your digestion. Best not to eat too much when travelling – you are usually sitting still for hours on end and not burning up calories.

● Travelling on planes is the perfect time to go on a juice or fruit fast. Not only is lots of liquid a good antidote to the dehydrating effects of cabin air but the presence of aeroplane 'unmentionables' only gives you instant willpower to see a fast through! Fasting also relieves the symptoms of jet lag when your internal clocks continue to operate hours behind or ahead of clock time at your destination. One of the things that slows down recovery from jet lag is eating meals at strange times. On west-east transatlantic flights for example, you are often served breakfast at 4am body time. By fasting you avoid this problem and arrive at your destination feeling light and lively instead of tired and suffering from indigestion. Make your first meal at your destination (be it breakfast, lunch or dinner) a light fruit or vegetable salad and take it from there.

● If you feel that you cannot possibly last the flight without something to eat, prepare a bag of fresh fruit, vegetables, sprouts and seeds, and a thermos of juice or herb tea or a bottle of spring water, to take with you. A juicy apple or orange way up above the clouds comes like a welcome breath of fresh air.

● If you are determined to tackle airplane food despite the preservatives and general tastelessness, you could try ordering a 'special' meal. You must notify the airline 24 hours before the flight and tell them the sort of diet you are on (salt-free, vegetarian, low cholesterol, etc.). Some airlines will even serve you a fresh fruit or vegetable salad! On long flights they can usually make arrangements to accommodate your needs. Special meals are often better than the ordinary ones, but don't expect a miracle!

When travelling abroad be sure to include as much fresh raw food as possible in your diet. We sometimes take seeds and pulses with us, along with plastic bags or a jar, and a strainer. It really is very easy to grow these little wonders in a hotel room, or even in bags while back-packing, and the benefits you reap – fresh goodies available for snacks day or night – more than make up for any inconvenience. Wherever you go eat a little of the local yoghurt. This will acclimatise your system to 'foreign' bacteria and your intestinal flora to coping with strange foods. Herb teabags are also a very good idea, infinitely preferable to tea and coffee. When you return from your travels you can cleanse your system with lots of fresh salads.

Occasional lapses and how to correct them

Having given all these hints about eating away from home, it must be pointed out that one 'different' meal every now and then is not going to harm you. By eating raw foods you strengthen your body, and a strong body can cope with even the most depleted foods . . . occasionally. But don't make it a habit or you may slip into the cooked convenience/junk food lethargy of not caring about yourself or what you feed your body. On a mainly raw diet one quickly develops a sense of what one's body will and will not tolerate and which foods are best avoided.

If you do find that a heavy meal or a series of heavy meals makes you feel out of sorts, use the Quick Corrector on pages 149–50 to put you right again.

Even on a mainly raw diet one can be thrown 'out of whack' by stress or by being obliged to eat a heavy meal. We scold ourselves when we overeat and curse when we come down with a cold the day before an important event. But it is very easy to put yourself back on the right track if you know how.

Most minor problems – hangovers, indigestion, fatigue, sluggishness, colds, muscle aches – are a question of system toxicity. Normal body functions are temporarily impaired by the presence of toxic wastes in the cells and blood-stream. There are five things you can do to eliminate these wastes.

Exercise   Going for a brisk walk or run helps you to elimin-ate toxins through your breath. By increasing the rate and depth of your breathing you give your cells extra oxygen to help them expel metabolic wastes.

Relax   Taking a short nap is also helpful. It gives your body a chance to restore itself undisturbed by other activities.

Do skin brushing   This encourages the elimination of wastes through the skin and stimulates lymphatic drainage.

Flush your system out   Drinking plenty of fluids (juices, herb teas) helps to wash toxins out of your body.

Return to raw foods and juices   What you eat and drink (as well as what you don't eat and drink) is probably the most important factor of all. Raw foods and fresh vegetable juices, as well as supplements of vitamin C, give your system the optimal conditions and energy for speedy recovery.

You can start the Quick Corrector plan at any time of day or night, adjusting the steps as appropriate. Acknowledge that you feel lousy and that there is a lot you can do to stop yourself feeling lousy. The occasional binge is often followed by cravings for foods which you know will not do you any good, chocolate for example. Recognise those cravings for what they are – a temporary imbalance in your system. The Quick Corrector assumes that you have come home after an evening of heavy food or too much to drink. Here it is.

Quick Corrector

Before going to bed
Take 1 or 2 g vitamin C in tablet or powder form (vitamin C removes toxic wastes from the system) and drink a cup of herb tea (peppermint or aniseed with a little honey).

On waking next morning
Drink the juice of an orange and half a lemon topped up with cold water, *or* the juice of one lemon topped up with hot water and sweetened with a little honey or molasses. Take another 1 or 2 g vitamin C.

During the morning
Go for a run or a walk of a mile or two – probably the last thing you'll feel like doing!

Do five minutes' skin brushing, followed by two or three alternating warm and cold showers.

Make some fresh vegetable juice – carrot and apple, carrot, apple and beet, or carrot, apple and celery – and drink this throughout the morning. If you don't have time to make juice, drink your favourite herb tea (hot or iced) instead.

If you get really hungry during the morning eat an apple or two, but it would be better not to eat anything until lunchtime because the liver is at its most busy removing toxins between midnight and noon.

Lunch

You should be feeling a lot better by now. If you are still feeling under par eat a bowl of grated apples sprinkled with orange juice and cinnamon. Add alfalfa or other sprouts to this if you have some available. If you are feeling well on the road to normal have a mixed raw vegetable and sprout salad instead, with an olive oil and lemon juice, or olive oil and cider vinegar, dressing. If you are at work and have to go out for a business lunch, order a plain vegetable salad (no cheese or meat of any kind) with a simple oil and vinegar dressing. Take another 1 or 2 g vitamin C.

During the afternoon

Drink juice or herb tea.

Dinner

Have a light salad *or* some steamed vegetables.

Before going to bed

Drink a cup of chamomile tea. You'll wake up the next day feeling re-energised and completely 'corrected'.

The Quick Corrector is not merely an emergency measure. You can use it as a once a week regime or incorporate parts of it into your everyday routine to keep you in tip-top shape and reduce the chances of succumbing to colds and other common ailments. It is also a very good way of preparing for important events when you want to be as clear-headed and full of energy as possible. Used once a week it is a way of gradually eliminating poisons which have accumulated over the years and getting your body used to a high-raw diet – a whole new way of living.

15 EQUIPPING A RAW FOOD KITCHEN

With a well organised, well equipped kitchen, raw food is a pleasure to prepare. But there is nothing more annoying than setting out to make a meal in someone else's kitchen and spending ages looking for a brush to scrub the vegetables only to find that the one you used was the floor brush! Let's look at some of the tools which are most useful for the raw food gourmet.

Go electric

Although it is nice to return to nature wherever possible, you have to draw the line somewhere. Using electric equipment takes the tediousness out of chopping vegetables, gives you a greater choice of textures, allows you to make splendid desserts, nut loaves, sauces, soups and whips, and cuts down enormously on preparation time. We find that a few simple machines give full rein to our imaginations. These are the raw chef's equivalent of the oven and the microwave. For those who like an 'all manual' kitchen we suggest alternatives, but they really are second best.

The three machines we consider vital are a food processor, a juicer and a blender, in that order. You can get by without a blender because a food processor does many of the same things, but it is useful nonetheless. There are appliances which combine the functions of all three, but having them separately allows you to work on several recipes at the same time and encourages helpers! As far as your pocket allows, choose good strong machines that will stand up to heavy use. If you have a large family it might be worth investing in catering or industrial models which are sturdier and can cope with larger quantities.

Food processors

A good food processor is a blessing to the raw food chef, there are so many remarkable attachments to choose from – a blade, several coarse to fine graters, various slicers and shredders. The blade attachment is excellent for grinding nuts and seeds, wheat and other sprouts, homogenising vegetables for soups and loaves, and making dressings, dips, and desserts such as ice cream. You can do most of these things with a blender, but if your ingredients are gooey they tend to stick around the blade and you spend ages scraping with very little to show for it. The blade in a food processor is removable and easy to scrape, so you lose very little. The grater, slicer and shredder attachments are terrific for making salads. With their help you can prepare a splendid Dish Salad (see page 219) in five minutes flat. Do experiment with all these attachments because, believe it or not, vegetables actually taste different depending on how they are cut up.

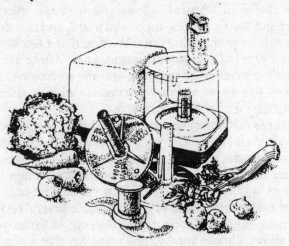

Juicers

The most important considerations here are power, capacity and ease of cleaning. The fewer fiddly parts there are to wash up the better. Some have a removable strip of plastic gauze in the pulp basket which is helpful in cleaning. Juicers come with various attachments depending on the make and model.

There are basically two types of juicer: the hydraulic press type and the centrifugal type. Some hydraulic presses are hand-operated and therefore less convenient than the electric kind, but some doctors who prescribe raw juices prefer them on the grounds that they reduce the amount of oxidation that takes place when juices are exposed to air. We have a centrifugal juicer ourselves. Centrifugal juicers are of two types: either they are separators, which operate without needing to be constantly cleaned out, or they are batch operators, which have to be cleaned out after every 2 lb

(roughly a kilo) of material has been juiced. That gives the separator kind the edge when it comes to convenience; they expel leftover pulp rather than keep filling up with it. But they tend not to extract juice as efficiently as the batch operator kind. If you decide on a batch juicer look for a large capacity model which does not require emptying too often. It can be infuriating working with a machine that insists on being cleaned out after juicing only two glasses when you are juicing for six people!

Blenders

There is not much to choose between blenders except their power. We recommend models of 400 watts or over (anything less will be unable to cope). Some have attachments for grating, chopping, kneading, etc. which are very useful. Glass models are preferable to plastic as plastic tends to stain and look tatty very quickly. Look for one that has a removable blade (the base unscrews) for ease of cleaning.

Other gadgets

Two other devices we find extremely useful are an electric citrus fruit juicer and a lettuce spin-drier. The citrus juicer has a central rotating cone onto which you press your halved grapefruits, oranges and lemons. Very quick and easy. There is nothing to stop you juicing citrus fruits in a centrifuge juicer, but you need to peel them first. The lettuce spin-drier is a great invention. There are several types, but our favourite is a basket which fits into a container with holes in the bottom and has a lid with a spinning cord. You put the whole contraption into the sink, put your lettuce or greens into the basket, put the lid on, run water slowly through the hole in the lid and pull the spinning cord. This spins the basket and expels the water, in theory cleaning and drying the greens. In practice they need to be rinsed before you put them into the basket, but by spinning you get beautifully crisp non-watery leaves very quickly.

Low technology alternatives!

The following are helpful if you cannot afford or have basic objections to electrical equipment. But you will be more limited in the number of textures and recipes you can prepare.

A sturdy grater – the box type with a fine, medium and coarse face, and a face for grating nutmeg and ginger

Hand coffee grinder for rendering down nuts, seeds and spices

Meat mincer – the sort you screw to the table, with coarse and fine cutters; good for grinding grains, seeds, nuts and sprouts

A strong stainless steel sieve for rubbing soft fruits through or extracting the juice from finely grated vegetables

Hand hydraulic juicer

A stainless steel 'mouli' rotary grater with coarse and fine grater inserts; quite effective for juicing finely grated fruit or vegetables

Pestle and mortar for grinding herbs, spices, flowers, etc.

A lemon squeezer

Wire salad basket, the sort you swing maniacally round your head in the garden

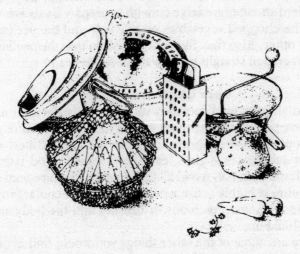

Knives and chopping boards

Of primary importance to raw food preparation are good knives and a good chopping board. At least two knives are essential, a large one for tackling spinach leaves, onions, carrot sticks and so on, and a smaller one for more delicate jobs. The best knives are made from carbon steel. Some enthusiasts disapprove of carbon steel because, unlike stainless steel, it encourages oxidation of cut surfaces, but we prefer them, for although stainless steel knives look nice they do not keep their edges as well and a sharp edge is important for creating beautiful salads. If none of your knives will cut a tomato without squashing it then they need sharpening! A good sharpener is worth investing in.

Good chopping boards are hard to find. Either they lose their pretty patterns with repeated chopping, or they warp when they get wet, or they are not large enough to slice an orange on without most of the juice running over the edge. Find a decent sized wooden board if you can, with runnels around the edge. Here is our solution to the problem. When we had a new kitchen installed we kept some big leftover pieces of formica-covered board. They make excellent chopping boards because it doesn't matter if they get scratched and they are large enough to prepare a salad on or leave the chopped vegetables on one side and the peelings on the other. Also they fit over the sink so that the peelings can be scraped straight into a waste bowl beneath.

Other basic equipment

All told, the raw food chef uses very few utensils – there are no heavy pots and pans to go in and out of the oven or to wash up! Choose dishes and platters made of inert or natural substances – glass, earthenware and wood rather than plastic or metal. Avoid all things made of aluminium as aluminium is highly active when it comes into contact with the acids in some raw foods. It can get into the body and slowly build up.

Here are some of the other things you would find in our kitchen.

A special 'vegetables only' scrubbing brush

A large colander, with feet so that it can stand in the sink to
  drain

For making yoghurt – *a pan to boil milk in* and various
  containers (glass jars, thermos flasks, earthenware
  pots)

Bread pans (preferably glass) for making vegetable loaves

Flat boards or trays for making sweet treats or Essene bread

Ice cube trays

A garlic press – achieves much better and quicker results than
  a pestle and mortar

Scissors for cutting up fresh herbs such as chives, parsley,
  mint, and so on

Salad bowls of different shapes and sizes

Soup plates, fairly wide and deep, for individual 'dish salads'

Salad platters – you can create attractive banquet-like effects
  by serving crudités arranged on a large platter, perhaps
  one with several compartments for dips

Several pairs of salad servers

A large pitcher for drinks, and a strainer

For sprouting

Although you can buy commercial sprouters we find home-
made ones simpler and less hassle. This is what you will
need.

Glass jars – the bigger the better

or Trays – baker's trays or the kind of trays that gardeners use
  for sowing seeds

Cheesecloth or metal gauze, and some rubber bands

Strainer

Dish strainer to stand sprouting jars in (optional)

Water purifier, because some sprouts insist on pure water
  before they will germinate. So might you if you are a city
  dweller and afflicted with bad tap water. In which case a
  water purifier can provide you with untainted drinking
  and water to use in sauces, soups and herb teas. Our
  favourite kind is in pitcher form and has a replaceable
  filter which needs to be changed every few months. You

regularly top it up with tap water so that you have a constant supply of pure water.

Storage matters

It is important to store living foods carefully so that they stay alive. We keep our seeds, pulses and grains in sealed polythene bags or airtight jars. Glass jars filled with beans and grains look particularly attractive and colourful. Empty sweet jars make useful storage containers, as do the plastic tubs in which one buys honey, peanut butter and margarine. Always cover salads with cling film (preferable to foil) as soon as you have prepared them, even if it is only for ten minutes while you prepare the rest of the meal, as this ensures minimal oxidation.

16 FRESH, FRESH, FRESH!

Fresh fruits and vegetables, and as many different varieties as possible, are the staples of a 75 per cent raw regime. All of them should be thoroughly washed in clean cool running water before use, not only those that have soil clinging to them or appear dirty or sticky, but also the spotless cellophane-wrapped supermarket variety. In fact these model specimens need special attention because they are often washed in detergent before packaging. Who wants vegetables tasting of washing-up water? Scrub anything that will stand up to a good scrubbing, using a brush marked Veg Only. Scrubbing is preferable to peeling since many of the valuable minerals and vitamins in fruits and vegetables are stored directly beneath their skins. A colander, one with feet, is useful for rinsing and draining all fresh produce, especially small items such as peas, mushrooms, radishes, sprouts and berries. We also find a salad spinner invaluable (see page 154). Never soak fruit or vegetables – it only dissolves the water-soluble vitamins out of them.

When choosing fresh fruits and vegetables, be fussy. If

the shopkeeper tries to palm you off with dowdy looking goods don't be afraid to say 'This cauliflower looks a little mangy. Could I have that one there instead?' Ask to sample grapes and things before choosing a particular kind; feel avocados and peaches; sniff melons and try pulling the top leaves out of pineapples to make sure they are ripe. If this does not endear you to the shopkeeper (and it seldom does) ask him to select something for you and if, when you get it home, you find that it is not ripe or good, take it back. It really is important to buy top quality produce. It makes all the difference between enjoyment and endurance.

Fruit and vegetable groups

Many people insist on combining specific groups of fruits and vegetables on the basis of their acid- or alkaline-forming properties. There is something to be said for this as different foods require different kinds of digestive enzymes and hence different pHs for optimal breakdown and use in the body, so that when eaten together some of the foods may not be fully digested. Even so the increase in enzyme activity that comes from eating a lot of raw food seems to enable the body to cope with a greater range of food combinations than on an ordinary cooked diet. Provided you chew everything well, so that your digestive enzymes are given the best possible chance to get to work on the food you swallow, you should have no worry about food combinations and digestive problems.

Here, in their approximate combining groups, are the fruits that appear in most greengrocers and supermarkets in due season:

acid fruits – *oranges, lemons, grapefruits, limes, tangerines, uglies, kumquats, strawberries, blackberries, gooseberries, raspberries, black, red and white currants, plums, damsons, cranberries, pineapples, pomegranates*

slightly acid fruits – *apples, pears, peaches, nectarines, cherries, pawpaws, mangoes, grapes, fresh figs, blueberries, kiwi fruit, lychees, passion fruit*

sweet fruits – *bananas, fresh dates, persimmons, most dried fruits; melons – watermelon, cantaloupe, honeydew, cassaba.*

Among the vegetables the combining groups roughly correspond to the part of the plant used:

leaves and stems – *lettuce (cos, Chinese, iceberg or Webb's Wonder, lamb's or corn salad, romaine, red or radicchio and Boston), cabbage (red, white, green), cress (watercress, mustard and cress, land cress), spinach, beetroot tops, turnip tops, spring greens, kale, Brussels sprouts, dandelion leaves, endives (curled, round leaved, French), celery, fennel, chicory*

roots – *carrots, beetroots, swedes, turnips, parsnips, radishes, horseradish, white radishes, celeriac, Jerusalem artichokes, kohl rabi, salsify, potatoes*

fruiting and flowering vegetables – *cauliflowers, avocados, broccoli, onions, leeks, spring onions, red onions, shallots, runner beans, French beans, stick beans, broad beans, peppers (red, green, yellow), pimentos, chillies, tomatoes, cucumbers, courgettes, aubergines, young marrows, squashes, pumpkins, asparagus, mushrooms and other fungi.*

It is especially important when switching to a mainly raw diet to delight the eye and the taste buds with as wide a variety of fruits and vegetables as possible. The act of discovering new tastes helps to make up for any sense of deprivation you feel at renouncing the bland homogenised flavours that food manufacturers insist on pushing at us. As far as possible buy fresh produce in season; that is when its vitamin, mineral, sugar and nutrient content is highest. Few fresh foods improve with keeping. Although the advice to seek out a shop that sells organically grown fruits and vegetables is fine in theory, most people buy their fresh produce in supermarkets and greengrocers and will continue to do so for practical and economic reasons. The surest method of obtaining unadulterated fruit and vegetables is to grow them yourself, although you cannot hope to emulate the variety available in the shops. An apple tree, a plum tree, and a walnut tree would be a good start; and radishes, spinach, cabbage and carrots are not too difficult to grow. Sprouts, of course, are the easiest, quickest, purest and most nutritious of all foods to grow.

Food for free

The annual parade of fruits and vegetables in the average greengrocer's shop gives you a very poor idea of the amazing variety of plants which are edible. The cooks of past centuries relied heavily on herbs, salad greens, nuts, fruits, seeds, spices, flowers and seaweeds gathered from the countryside. Today many of these wild plants occur fairly locally.

By their very nature 'weeds' possess incredible strength and resistance to all that weather, poor soil, parasites and disease can throw at them, and remarkable reproductive capacities. They have strong foraging roots which penetrate the soil to great depths and growing shoots that can struggle up to the light even through concrete. They are full of vitamins and minerals. Most should be used in moderation as they have sharply individual tastes.

Certain parts of the majority of wayside plants are edible, in small quantities, but a few are extremely poisonous. We recommend that you consult a wild plant encyclopedia before you start foraging, one that gives all the fascinating culinary and medicinal uses of our native plants. In the meantime, here are a few you might like to try.

Young dandelion leaves These contain lots of nutrients, including calcium, potassium, sodium, silicon, phosphorus, iron, oxalic acid (cleanses the gall bladder and kidneys), vitamin A and C, and choline (a substance essential to the efficient conduction of nerve impulses). The youngest leaves are delicious in salads. Be sure to pick real dandelions (there are many frauds): they have no hairs, deeply toothed edges, are soft and not glossy, and the flowers are borne singly on a single stem. Dandelions are notably alkaline and therefore good for arthritis sufferers and anyone addicted to alcohol or white bread! They are digested very quickly and have a tonic action on the liver and kidneys.

Dock and sorrel These have a good mineral balance and are particularly high in sulphur. They are good body cleansers, have a tonic effect on the blood, and due to their vitamin A content are good for an overworked liver. Sorrel especially is delicious in salads and soups, giving a very fresh lemony taste.

Horehound   This is high in vitamin C. Chopped and mixed with a little honey it is a good cold remedy.

Plantain   A very common weed this, and noted for its blood cleansing properties. It is high in chlorophyll and can be juiced with other vegetables to give a blood-boosting, chlorophyll-rich drink.

Chickweed   After bindweed the intruder most hated by gardeners. Has tiny starry white flowers and grows straggling across the ground. Its pale stems and small soft green leaves make a nice addition to green and herb salads. Don't confuse it with milkweed or mouse-ear chickweed.

Couch grass   A good diuretic this, particularly for women who retain water at the beginning of their periods. The runners can be chopped into salads.

Purslane   Highly alkaline, purslane has a soothing effect, especially on over-acid stomachs. Its bright green fleshy leaves, borne on purplish stems, can be shredded and added to salads.

Just to give you an idea of the variety awaiting the discerning rambler here are some other palatable and flavourful free foods:

for salads – *salad burnet, watercress (usually an escape), very young lime, beech and hawthorn leaves, fat hen, burdock, marsh samphire, common mallow leaves, chicory, wild cabbage*

herbs – *water mint, spearmint, apple mint, pennyroyal, wild marjoram and thyme, bay, jack by the hedge, ramsons (wild garlic), lovage, chervil, borage*

seeds – *fennel, coriander, juniper*

as flavourings or infusions – *tansy, sweet cicely, galingale*

roots – *horseradish, pignuts, wild parsnip*

seaweeds – *kelp, dulse, laver, carragheen*

flowers, as flavourings or for adding to salads – *sweet violets, honeysuckle, wild rose petals, red clover,*

chamomile, nasturtiums, broom buds, elder flowers, meadowsweet
nuts – *cobnuts (filberts), sweet chestnuts, beechnuts*
fruits – *blackberries, wild* pears, crab apples, rose hips, sloes, elderberries
fungi – *field mushrooms, horse mushrooms, ceps, boletus, horn of plenty, shaggy inkcaps, blewits, chanterelles.*

It is interesting that weeds growing in areas where the atmosphere is polluted seem not to be affected by their unhealthy environment. This is because most weeds have two root systems, one superficial in contact with the air and soil surface, and the other deep underground where all the nutrients are. The nutrients absorbed by deep tap roots pass up into the main body of the plant for synthesis into compounds which the plant needs. Pollution of ground water is different. The less cultivated the area where you pick your weeds the better, as it is less likely to contain pesticide residues, artificial fertilisers or dissolved detergents.

We are not advocating a return to a hunter-gatherer existence or that you ransack the countryside for these plants. But wild greens and herbs are generally more nutritious and healthy than the cultivated varieties you buy in shops. They also make one realise to what extent cultivation and mass production techniques have muted the taste experiences available to us. Every single cultivated plant was once a wild species. We feel, quite simply, that some curiosity about wild plants, and a respect for them, is part of a Raw Energy lifestyle.

Some raw food cautions

Some raw foods, especially the pulses, have a rather chequered image among health food enthusiasts. Much depends on whether you eat them raw, cooked or sprouted, and on complementary nutrients in your diet.

Some of the beans, for instance, have certain negative attributes. Soya beans, broad beans and red kidney beans contain a trypsin inhibitor, a substance which blocks the action of some of the enzymes in the body which break down protein. This means that a proportion of the valuable amino acids they contain cannot be used. Many years ago researchers discovered that soya beans would not support life unless they were cooked for several hours. Cooking and sprouting neutralise the trypsin inhibitor.

In fact sprouting greatly improves the safety and nutritional quality of all pulses, seeds and grains. The enzymes which go into action during germination not only neutralise trypsin inhibiting factors but also destroy harmful substances such as phytic acid, an important constituent of grains, which tends to bind minerals making them unavailable to the body. Sprouting, by destroying phytic acid, releases the minerals for use.

Chick peas also contain a trypsin inhibitor, rendered harmless by sprouting but not by cooking. Green peas contain a hemagglutin which resists cooking (hemagglutins inhibit growth by combining with the cells lining the intenstine and blocking nutrient absorption) but in such small quantities that one would have to eat virtually nothing but peas, raw or cooked, to experience any adverse effects. Eating raw lima beans has been known to cause death.

Everyone knows that rhubarb leaves are poisonous. This is because they contain massive amounts of oxalic acid. The stems contain much less. Swiss chard, beet greens, turnip and mustard greens, collards, kale, spinach and French sorrel also contain oxalic acid. If eaten to excess the acid in them can block calcium absorption and cause kidney damage as acid crystals deposit themselves in the urinary system. Although oxalic acid is not destroyed by cooking, it appears qualitatively different from the oxalic acid in raw foods. Norman W. Walker, the American raw food expert, insists that the oxalic acid in raw foods has no harmful effect. On the contrary it stimulates peristalsis (rhythmic squeezing of food through the gut). He and others often recommend adding spinach juice to other fresh vegetable juices.

Another group of vegetables, the brassicas, has been blamed for suppressing thyroid activity. Cabbage, Chinese leaves, watercress, kale, turnips, Brussels sprouts and mustard all belong to the genus *Brassica*. They contain compounds called thioglucosides which can disrupt the function of the thyroid gland and have been shown to contribute to the development of goitre. Drinking the milk of animals who have been grazing on these plants can also disrupt thyroid function. But you are unlikely to experience the undesirable effects of the thioglucosides if you take in adequate iodine (from fish or seaweed) in the diet. It is only people deficient in iodine who suffer them.

People are often cautioned against eating raw eggs. Raw egg yolks are fine but the whites contain avidin, a substance which combines with the B vitamin biotin and prevents its absorption into the blood. One young man who ate lots of raw egg whites developed scaly skin, anaemia, anorexia, nausea and muscle pains because of biotin deficiency. This is of course an extreme case. Raw egg whites are less likely to cause such symptoms if eaten with the yolks. Also, the albumen in egg whites can easily enter the bloodstream undigested and can cause allergies. We generally use only yolks in our cuisine. However we do not think twice about preparing a delicious egg nog occasionally, using the yolks and the whites. It is simply a question of moderation.

17 STOCKING YOUR LARDER

Making raw food a part of your life means making changes. These changes begin with what you choose to put in your larder. Take a look at the contents of your kitchen shelves and cupboards and ask yourself 'Do I really feel good about putting those things into my body?' Chances are bottles of squash, tins of ravioli and bags of white sugar and flour will force you to say 'No!'

But before you can begin to stock up with healthful replacements and make your larder a place to be proud of, you have to clear away old habits. The best way to do this is gradually. You will probably find that after feeling the first benefits of raw food you want to get rid of all your 'junk' food *now*. It is a good idea to put all suspect foods in a box in the corner and see just how little you want to use them. Then if you really want to be free of them you can donate them to your favourite charity.

We find it very annoying to be given lists of rigid do's and don'ts – things one must and must not eat or drink. Really an awareness of what is good and bad must come from you personally, and that awareness gets stronger the more of your food you eat raw. In the meantime remember that the further a food gets from its natural state the worse it is for you. Anything artificial or chemically processed is automatically suspect. If the origin of what you are eating or drinking is lost in the mists of processing, avoid it. Here are some of the main offenders: most convenience foods; frozen prepared foods; white flour and white bread, even packaged 'brown bread'; white and brown sugars (not dark brown molasses sugar); coffee and tea; anything containing sugar (saccharin, glucose syrup, corn syrup, dextrose, etc.), preservatives, flavourings, permitted colour, emulsifiers, edible starch, stabilisers . . .

Become an avid label reader (if you dare!) and look out for all those additives – the longer the names get the more dangerous they may be!

Seduce your taste buds

Variety is the secret of a really useful raw food larder because at certain times of the year you will be limited by the fresh vegetables and fruits available. The foods which provide this precious variety and make raw food exciting at any time of year are: various grains; seeds and pulses for sprouting; seeds; nuts; dried fruit, for those times when fresh fruit is hard to come by; herbs, seasonings and spices.

Eating a raw diet heightens your senses, which is especially important in appreciating good food. You begin to notice and delight in the hundreds of delicate aromas and taste sensations which exist in foods in their natural state but which are lost in cooking. A well-flavoured casserole is not one plastered in tomato ketchup, stock cubes, salt and pepper. How many wonderful herbs and spices are there if one only takes the time to become acquainted with them. In olden times of course herbs and spices were used to smother the taste of less-than-fresh meats and other foods. But properly used, seasonings should augment rather than overpower the taste of food. Used sparingly they give endless subtle flavour combinations.

In the next few pages we will show you many of the seasonings that we use in our kitchen. You will notice that some of the condiments suggested are cooked. This is no accident, for we see no reason to be fanatical about not using some lightly toasted seeds or dry-fried spices to heighten the flavour of a dish. You will also notice that we have omitted salt and suggested various salt alternatives. This is because most people, without even realising it, take far too much salt. Excess salt is one of the culprits in high blood pressure and upsets the body's sodium/potassium balance. Try to broaden your knowledge and use of herbs and spices. You can build up a handsome selection of seasonings by buying a new one each time you go shopping and growing fresh herbs in window boxes or in your garden. You'll be surprised at the difference they make!

Herbs and spices

Remember to buy spices and dried herbs in small quantities from shops which sell a lot of them, as they soon lose their freshness. The most flavourful herbs are freeze-dried, but these are hard to find. Vacuum-sealed herbs are a good bet. Herbs and spices should be stored in the dark as the light affects their potency (so much for herb and spice racks!). For the sake of flavour it is best to buy whole spices and grind them as needed in a coffee grinder or with a pestle and mortar. If you have never ground your own spices you'll be amazed at their exhilarating aromas and tastes. With fresh herbs, wash and dry them well, then store in sealed polythene bags in the salad compartment of your fridge.

Here are some of the many seasonings which feature in the Recipe Section which begins on page 205.

'Sweet' spices (for fruit salads and dressings, desserts and ethnic dishes): allspice, angelica, aniseed, cardamom, cinnamon, cloves, coriander, ginger, nutmeg and mace.

'Savoury' seasonings (for salads, soups and raw main dishes): basil, bay, caraway seeds, cayenne pepper, celery seeds, chervil, chilli powder, chives, coriander leaves (fresh), cumin, curry powder, dill, fennel, garlic, horseradish, juniper berries, kelp, lemon balm, lovage, marjoram, mint, mustard seeds, onion, oregano, paprika, parsley, pepper, poppy seeds, rosemary, sage, savory, sorrel, tarragon, thyme.

This list, though not exhaustive, is somewhat overwhelming! The ready reference chart on pages 170–76 describes the 17 herbs and spices we find most useful, and certainly those worth becoming acquainted with if you are not already.

Savoury seasoning powders

We find the following seasonings useful in those moments of panic when a dish tastes lacking in that vital 'something'. One rescue remedy we wouldn't be without is a wonderful vegetable bouillon powder made by Marigold Health Products and called simply Swiss Bouillon Powder. We use it in dressings, sauces, ferments and soups, and in seed and nut dishes. You can of course make your own seasoning powder. A recipe we are particularly fond of is made up as follows: 1½oz or 50g onion powder, ½oz or 15g garlic powder, 2oz or 70g celery leaf/comfrey powder, 1 teaspoon cayenne pepper, 1 tablespoon kelp, ½oz or 15g ginger powder. You can also make your own spice mixture for drinks and desserts by combining powdered allspice, coriander seeds, cinnamon, cloves, ginger, nutmeg and mace to taste.

Condiments and other flavourings

Mustard You can buy mustard as seeds or in dry or paste form. The dry powder is sometimes useful in dressings. So are many prepared mustards. The most delicious mustards are French. They are milder and more aromatic than English mustards. Moutarde de Meaux, fine and full grain, is particularly delicious and makes a palate-tickling addition to all sorts of salad dressings. Dijon and Bordeaux are very flavoursome too.

Tahini This is a paste made from finely ground roasted or unroasted sesame seeds. Tahini has many uses: in tahini mayonnaises, added to seed and nut dishes, or mixed with honey as a topping for fruit and desserts.

Tamari This is a type of soy sauce made from fermented soy beans and preferable to soy sauce because it contains no wheat. Unfortunately it does contain sea salt, so it should be used in moderation. Nevertheless, it gives an authentic Chinese taste to dishes and is good for livening up bland dressings or sauces.

Yeast extract This can be used as a substitute for vegetable bouillon. It is rich in B vitamins, but very salty, so it too should be used sparingly.

Vinegar Apple cider vinegar is the best as it contains malic acid, an aid to digestion. All other vinegars, except naturally fermented wine vinegar, contain acetic acid and should be avoided as acetic acid destroys red blood cells and interferes with the digestion and assimilation of food. It is also known to contribute to cirrhosis of the liver and ulcers.

Vanilla Try to find real vanilla essence rather than the more common synthetic vanilla flavouring. Used in nut milks or yoghurt drinks and desserts vanilla gives a deliciously round, mellow flavour.

Flower waters Orange flower water and rose water contain the essential oils of their respective flowers. They can be bought in delicatessens, Middle Eastern food shops, herbalists, some health food shops and also at the chemist's. They make delicious additions to fruit salads and sweet treats.

HERBS AND SPICES CHART

Seasoning	Description/ Part used	Used for	Tips
Allspice	The dried fruits of an evergreen. They resemble peppercorns, dark red/brown in colour.	Tastes like a combination of spices: cinnamon, juniper berries, cloves, nutmeg. Good for fruit and yoghurt drinks, herb teas, fruit desserts, cakes, puddings, mueslis, marinades.	The whole fruits can be bought and ground in a pepper grinder so that the spice is really fresh.
Basil (sweet)	The leaves of the plant, but also the flowering tops.	Has a sweet, spicy, slightly peppery taste. Delicious fresh in salads or in salad dressings. Goes particularly well with tomatoes and in Italian and Greek dishes.	Basil is so much better fresh! Grow it in pots on your kitchen window sill. It takes up very little space (especially the dwarf variety) and is said to keep flies away.
Cayenne pepper	The fruits of a variety of tree from the Capsicum group of which chillies, peppers and paprika are also members.	Comes in powder form and is a dull red colour. A very hot spice which is used to add bite to dishes. Particularly good in seafood and Mexican recipes. Can be added to bland dressings to give zip.	Can be a pepper substitute, for a change, but use sparingly. It is helpful as an appetite stimulant and digestive. For a gentler flavour see paprika, cayenne's sister spice.

Chives	Belong to the onion family but have no large bulbs. The green grass-like stems are used.	Have a much milder, fresher taste than onions and a quality all their own. Particularly good in dips and cheeses, and attractive used to garnish soups and vegetable cocktails.	Another easy herb to grow. The subtle flavour is mostly lost when dried – dried chives, although attractive sprinkled over dishes, have a bland flavour. Use scissors to chop the fresh herb finely as a sprinkling for soups, dips and sauces.
Cinnamon	The inner bark of a species of laurel. Bought in stick or powder form.	Has a warm, sweet aromatic flavour. The sticks are great in hot drinks and can be given to children to chew on. Used to spice desserts as well as sauces, tomato ketchup for example.	Cinnamon has antiseptic properties and is supposed to be particularly good for soothing an upset stomach.
Dill	The fruits or seeds of the plant are used, as well as the feathery leaves.	Has a penetrating, slightly aniseed, slightly citrus flavour. It is especially delicious with cucumber and cauliflower and is used in pickling. It also gives a refreshing flavour to seafoods and garden vegetables.	A particularly good digestive, dill has been used in the past to stimulate the appetite and settle the stomach and even as an aphrodisiac. Chewing the seeds is said to sweeten the breath.

Seasoning	Description/ Part used	Used for	Tips
Garlic	The bulb of the plant, consisting of numerous cloves, is used. Dried garlic powder is also available, but steer clear of garlic salt.	A pungent, strongly flavoured herb – a little is often enough to give a dish or dressing that little something that it lacks. Gives the delicious French-cuisine aroma to food. Especially useful in salad dressings.	Many people find garlic on the breath unpleasant. A little parsley chewed after a strong garlic dish will cure this. If you find garlic dressing too much to stomach, try rubbing a peeled garlic clove around the inside of the bowl before the vegetables go in. This will give a faint suggestion of its presence without any overwhelming taste. Garlic is one of nature's best antiseptics.
Ginger	The stem or rhizome of the ginger plant. Bought fresh, dried or powdered.	A hot, pungent spicy flavour, used in biscuits, to make refreshing drinks, in chutneys, sprinkled lightly on melon, or on root vegetables like carrots and parsnips. Adds 'warmth'.	Fresh ginger is very strong, so use it sparingly. Acts as an aid to digestion by stimulating the secretion of digestive enzymes.

Kelp	This is a type of seaweed which can be bought in powder form.	Used as salt substitute and put in the salt shaker on the table at meals. Useful in sauces and dressings.	Kelp is one of the most nutritious foods available. Unlike salt, kelp has a perfect mineral balance. It contains many trace minerals and is particularly high in iodine, which is helpful for obesity since it stimulates the thyroid gland to increase metabolic rate.
Marjoram	There are different types, but the best known is wild marjoram which is the same thing as oregano. The leaves are used.	A delicious spicy, slightly pungent herb. Used in almost all ready-mixed herb combinations that can be bought. It is used particularly in Greek cookery and combines well with other herbs such as thyme. Good in vegetable soups, with courgettes, and in Italian dressings.	Another good digestive and also mildly antiseptic.

Seasoning	Description/ Part used	Used for	Tips
Mint	The leaves are used. Different types include: pepper-, spear-, orange-, apple-, pineapple-, cat-, ginger-, and even eau de cologne mint!	The flavour of mint is better known than that of any other herb because it is so widely used in making sweets. However, the different varieties are well worth trying and cultivating yourself as each has its own subtle flavour tones. Mint is delicious in drinks, in salads, and even in fruit salads where it gives a really fresh flavour, and in some savoury dishes. The leaves are good for garnishing.	Particularly good as a stomach settler, the fresh leaves can be chewed on directly or infused to make an after-dinner tea.
Paprika	Bought in powder form. Rich red in colour (see cayenne pepper).	Paprika has a mild, slightly sweet, 'dull' flavour. It should be bought in small amounts so that it is fresh. It is especially good dusted sparingly on pale vegetables, soups and sauces because of the beautiful colour it imparts. It adds a subtle spiciness to dressings.	Can be used, like cayenne pepper, as an attractive substitute for white or black pepper.

Parsley	The leaves and stems are used.	Parsley has a rich 'green' flavour compatible with many other herbs. It is good chopped in patties, nut loaves, green salads and dressings, and as a garnish for almost any dish. Plain-leaved or common parsley has the best flavour; curly or moss parsley is better for garnishing.	Rich in vitamins and minerals, parsley has many medicinal properties. In small amounts parsley juice is said to improve the blood transport system and to be of value in treating kidney problems.
Rosemary	A shrub. The leaves are used.	Highly aromatic, tasting slightly of pine, rosemary has a pleasant but strong flavour and is best used in moderation. It is good in cheeses and marinades. Sometimes used in fruit dishes, especially apple.	The scent of rosemary is often used as a base for perfumes and bath oils. It is supposed to be good for clearing headaches and improving the circulation.

Seasoning	Description/ Part used	Used for	Tips
Sage	A shrub. The leaves are used.	Has a strong individual flavour, often included in dried herb mixtures. It has a particular affinity for onions and is good in savoury nut dishes. Adds flavour to seed and nut ferments and can be used sparingly in creamed soups.	Sage tea is a well known gargle for sore throats, while the leaves rubbed on the teeth and gums cleanse and strengthen them. Useful to help digest rich or fatty foods.
Thyme	A shrub. Both leaves and flowers are used.	Has a wonderful warming, sweet but full flavour. One of the main ingredients of *bouquets garnis*. Delicious with courgettes and bell peppers. Adds an aromatic sharpness to seed and nut dishes and goes particularly well with oily fish.	Has many medicinal applications. The Romans and Egyptians used it as an appetite stimulant and a digestive. It is also said to stimulate the intelligence!

Sweeteners

Honey This makes a wonderful substitute for sugar in drinks and desserts. There are dozens of different honeys ranging from mild flowery ones such as acacia and orange blossom to very pungent ones such as Mexican or Tupolo pine honey. Clear honeys are best for drinks and set honeys are nice on breads. The best honeys are those labelled 'organic', the product of bees foraging for flower nectars. Many commercial honeys are made by bees fed directly on sugar. Honey contains many useful trace elements and is an easily digested form of energy.

Molasses This is one of the super foods. It is the bulk that is left over from sugar refining and is as good for you as sugar is bad. It contains all the minerals and vitamins, particularly the B vitamins, that are taken out of the refined product. Some molasses is quite overpowering in its flavour and tends to have an unpleasant sulphur tang; this is because sulphur has been used in the refining process. Unsulphured molasses, however, is quite heavenly and can be eaten straight off the spoon. If kept in the refrigerator it becomes firm and thick and is wonderful on muesli, in yoghurt, or spread on bread.

The best kind of sugar is raw cane or molasses sugar, which is unrefined. But with honey and molasses in your store cupboard you do not really need to use sugar and are better off without it. Also, when you eliminate concentrated sweet foods from your diet you find you don't really want them any more.

Dried fruits Finely chopped dried fruits soaked in orange or lemon juice are wonderful sweeteners in desserts, mueslis and breads. Look for those that have not been treated with sulphur during the drying process.

Oils and fats

Buying good oil is very important. The best oils are the fresh unrefined cold-pressed sort. This is because they are extracted from the raw seeds under mechanical pressure rather than by heat and chemical processes. Cold-pressed oils contain essential fatty acids in a form that your body can use. In many heat-processed oils these usable 'cis' fatty acids have been chemically changed into 'trans' fatty acids which can not only be actively harmful but can also prevent your body using the 'cis' fatty acids in the rest of your food. Olive oil is good and adds a distinctive flavour to salad dressings. It is quite heavy though and some people prefer a lighter oil, such as sunflower and sesame, both delicious. Most corn and safflower oils are mechanically extracted, but using steam as well, and so are second best. Walnut oil, if you can get it, is wonderful but expensive.

Nuts

When buying nuts make sure they are really fresh. The rancid oils in old nuts are harmful to the stomach, retard the secretion of pancreatic enzymes and destroy vitamins. If nuts are fresh and still in their shells you can buy them in largish quantities and keep them in an airtight container in a cool dry place (best in your refrigerator) for up to several months. Shelled nuts should be bought in much smaller quantities and they too should be refrigerated. It is a good idea to buy several different kinds and mix them in your recipes so that you get a good balance of essential amino acids. These are some of the nuts we regularly stock up with: almonds, brazils, cashews, coconut (fresh or dessicated), hazelnuts, macadamia nuts, peanuts (strictly speaking a legume), pecans (similar to walnuts, but less bitter), pine kernels, pistachios, tiger nuts and walnuts.

Seeds

Again be sure to buy really fresh seeds from a shop that sells a lot of them. The three seeds which provide an ideal combination of protein and essential fatty acids are: sunflower, pumpkin and sesame. Other seeds worth trying, mainly for seasoning, are: poppy, celery, caraway, dill, fennel and aniseed. The last four, separately or together, make good 'nibbles' between meals. In the eighteenth and nineteenth centuries people used to carry a mixture of these seeds in their pockets and munch them to keep their appetites at bay.

Grains

These figure in many of our bread, muesli and savoury recipes and are best used sprouted. Those we use most often are: wheat, rye, tricitale (a wheat/rye hybrid), oats, barley (groats) and millet. Occasionally we use them unsprouted but soaked for muesli: wheat and rye flakes, rolled oats, barley kernels and couscous. Millet is a particularly good grain and worth getting to know because it is the only alkaline grain and the only one that contains all eight essential amino acids. It can be used sprouted or soaked. Buckwheat, often classified as a grain, is in fact a member of the *Polygonaceae* family which includes rhubarb and sorrel, and it is the triangular seeds of the plant that are sold. They are often pre-roasted, but can be bought raw. As with millet they can be eaten soaked or sprouted. The sprouts should be about 4 cm or 1.5 in long before they are eaten. Buckwheat (which in truth is not a grain at all but is cooked like one) is particularly good for you because it contains rutin which appears to have an uplifting effect on the spirits and is one of the principal nutritional factors used to treat atherosclerosis (hardened arteries). It is also very good for people who bruise easily and have fine broken veins under the skin.

Pulses, seeds and grains for sprouting

Remember, all whole pulses, seeds and grains are living, breathing (yes!) things and should be stored and handled with care. There is a lot to say about this wonderful category of raw foods and we say most of it in the next chapter, *D.I.Y. sprouts, yoghurts and cheeses.* To start your sprouting career here are some you might like to stock up with, because they are the easiest to grow: *aduki beans, mung beans, soya beans, lentils, chick peas; alfalfa seeds, fenugreek seeds* and *radish seeds; wheat.*

Dried fruits

These deserve a place in your store cupboard in their own right, not merely as sweeteners. They do not need to be cooked. Twelve or eighteen hours' soaking in the minimum amount of water in a warm place will plump them up just fine. When buying them be sure they have not been dipped in glucose (figs, papaya and banana chips often have been) and that they have not been sulphur-dried. The best kind have been dried naturally, in the sun. Where possible look for whole or half fruits rather than pieces. You have a choice of: *raisins, sultanas, currants, apricots* (look out for the un-stoned Hunza variety), *peaches, prunes, pears, figs, pineapples, papaya, dates* and *bananas*.

Beverages

Herb teas or tisanes We drink these as substitutes for tea and coffee. Decaffeinated coffee can be almost as bad for you as coffee itself because of the chemicals used to remove the caffeine. But why not try *dandelion coffee* or a *barley and chicory* beverage? You will find a good selection of delicious beverages in any self-respecting health food shop. Then there are *fresh fruit and vegetable juices*. Make your own if you can. Second best is to buy packaged, bottled or canned juices, but scrutinise the labelling for the words 'pure' and 'free from additives, sweeteners and preservatives'. You can also buy fruit concentrates to which you add water; these are made purely from fruit, but they are 'cooked'. Something you may not have heard of is *rejuvelac*. This is the soak water from wheat sprouts. Made properly it is sweet-tasting and has many health benefits. Lastly, a word about the basis of all life: water. A water purifier (see page 157) may be a good idea if your tap water has a taint to it. *French bottled spring water* is quite safe and often has a good helping of minerals in it.

Dairy products

The best dairy products are those made from goat's milk. Goat's milk is more digestible than cow's because its protein and fat molecules are much more finely divided. A nutritious source of raw protein, goat's milk is also rich in the essential fatty acids and has an excellent balance of minerals. Even people allergic to cow's milk (and there are many) tolerate small amounts of goat's milk. *Goat's milk cheeses* and particularly *goat's milk yoghurt* are delicious and highly beneficial. The bacteria in yoghurt do a great job raising the tone of your intestinal flora! Cow's milk cheeses and yoghurt are all right occasionally, unless you have an allergy to cow's milk, and you will find that eating a high-raw diet tends to decrease allergic reactions.

Special foods

Carob (St John's bread)   This is sold as a powder or flour and is made from the pods of an evergreen Mediterranean tree. It is a superb chocolate substitute, and good for you too. Unlike chocolate it does not contain caffeine. Instead it is packed with minerals – calcium, phosphorus, iron, potassium, magnesium and silicon – and with the B vitamins B_1, B_2 and niacin, a little vitamin A and some protein. Carob powder is often sold toasted, but try to find the untoasted variety, which is lighter in colour. It can be bought from most health food stores. We use it to make chocolate drinks, desserts and sweet treats.

Seaweeds   These really are worth trying! Seaweed tastes better than it looks. Some varieties are positively mouthwatering. In fact seaweed is the most nutritious form of vegetation on this planet. It has as many as 41 trace elements. Its high iodine content is good for the thyroid gland, particularly if you are overweight and suffer from a low metabolic rate. It contains almost the whole alphabet of vitamins – A, B, C, D, E and K – and also a substance called alginic acid which speeds up the elimination of toxins and even reduces the amount of atomic radiation absorbed by the body. You can buy it or harvest it straight from the beach! Some of the varieties you will find shrink-wrapped in health food shops are: *arame* and *hijiki*, especially rich in iron, and a decorative addition to any salad; *dulse*, one of the nicest tasting, deep red in colour with a slightly spicy taste – can be chewed as it is or soaked and added to salads; *laver*, used with oats to make the famous Welsh laver bread, and soft in consistency – has a higher protein content than any other variety; *nori*, comes in thin dark purple sheets and makes a convenient wrapping, vine leaf style, for salads or thick dip fillings; *kelp*, especially rich in iodine and used in powder form as a flavouring; *kombu*, usually cooked in soy sauce to make it tender; *agar agar* and *carragheen*, both used as an alternative to animal gelatin to make sweets, jellies and aspic – agar agar flakes, unlike agar agar powder, have not been chemically processed.

Tofu   This is soya bean curd made from soya milk. It is the staple protein of at least one billion people throughout the Far East. All soya bean products are very high in protein and low in carbohydrate and therefore much favoured by dieters. Soya beans also contain a high amount of lecithin which not only fights the build-up of cholesterol deposits but is also essential for proper functioning of the brain and nervous system. Both soya milk and tofu can be found in health food stores. They have a very delicate, slightly nutty taste and are very versatile. One disadvantage is that unless they are cooked or made from sprouted beans they contain phytic acid, which binds zinc and other important minerals. We use uncooked tofu in moderation, or wok-fry slices of it and flavour with a little soy sauce.

Wheatgerm   This is a delicious and valuable food high in vitamins E and B. Like molasses it is what is left when the original raw food, in this case wheat, has been refined. It should be bought raw rather than toasted, and refrigerated to keep it fresh. We sprinkle it liberally on salads, mueslis and desserts.

18 D.I.Y. SPROUTS, YOGHURTS AND CHEESES

Home-grown sprouts and home-made yoghurts and other ferments are the purest foods one could possibly eat and the richest in digestion-promoting enzymes, and you will find many original uses for them in the Recipe Section which begins on page 205. In this chapter we tell you how to go about ensuring your own supply of these excellent foods. There is no need to rush out and buy special equipment, although you do need a warm place – an airing cupboard or a shelf near a radiator – where you can put your trays, jars and pots. Sprout-growing and yoghurt-making are a wonderful way to introduce children to the idea that the best foods are the result of natural processes, not those that come straight from the supermarket shelf or freezer cabinet.

A guide to sprouting

Sprouts can be grown by anyone, anytime, anywhere. Some hiker friends of ours even take packets of seeds with them and sprout them in their packs as a fresh, chlorophyll-rich addition to their camp food. All you need to start your own indoor germinating factory are a few old jars, some pure water, fresh seeds/beans/grains and somewhere warm. The Sprouting Chart on page 186 lists all the different sprouts with soak times, growing times, quantities and useful hints.

An improvised sprouter can be anything from a plastic bowl to a polythene bag, but the best sort is a wide-mouthed glass jar, or rather several wide-mouthed glass jars. Some people like to neatly cover their jars with cheesecloth, or nylon or wire mesh, secured with a rubber band. If you are very fussy you could use bottling jars and hold the cheesecloth in place with the screw-on rims. But the simplest way is to use open jars and cover them with a tea towel to stop dust or insects getting in.

For argument's sake let us presume that you are going to sprout some mung beans – the procedure is of course similar for all pulses, seeds and grains. Remember that a handful of seeds, beans or grains will give you approximately eight handfuls of sprouts. This is what you do.

● Put the beans in a large sieve, remove stones, other foreign material and broken beans and rinse well under the tap.

● Put the beans in a jar and cover them with a few inches of pure water (although rinsing can be done with tap water, the initial soak, during which the beans absorb a lot of water to set their enzymes working, is best done in spring/filtered/boiled water as the chlorine in tap water can inhibit germination).

● Leave the beans to soak overnight, or for about 15 hours, in a warm place (soak time varies according to the beans/seeds/grains one is sprouting).

● Pour off the soak water. If none remains you still have thirsty beans on your hands; give them more water to absorb. The soak water is good for watering house plants. Some people like to use it in soups or drink it, but we find it extremely bitter. Also the soak water from some beans and grains contains phytates, whose bactericidal properties protect the embryo plants from being invaded by microbes in the soil. Phytates also interfere with certain biological functions in humans, including the absorption of minerals such as zinc, magnesium and calcium, and are therefore best avoided.

● Rinse the beans either by pouring water through the cheesecloth top, swilling it around and pouring it off several times, or by tipping the beans out into a large sieve and rinsing them well under the tap before putting them back in the jar. Be sure that they are well-drained because excess water at this stage may cause them to rot. If you are using a cheesecloth-covered jar leave it tilted in a dish drainer to allow all the water to run out. Repeat this rinsing and draining morning and night. If the weather is very hot your beans may need a mid-day rinse too.

- Let the sprouter stand in a warm place between rinsings. Sprouts grow fastest without light and in a temperature of about 70°F.
- After about 3–5 days your mung beans will have sprouted and will be ready for a dose of light and sunshine. Put them on a sunny window sill. Keep them moist (a plant spray is ideal for this) and make sure they don't get too hot and start to roast.
- After a few hours in the sun most sprouts are ready to eat. Optimum vitamin content occurs 50–96 hours after germination begins. They should be rinsed and eaten straight away or stored in the refrigerator in an airtight container or sealed polythene bag. Some people dislike the taste of seed hulls – those of mung beans are a little bitter. To remove them simply place the sprouts in a bowl, cover with water, stir gently and the seed hulls will float to the top. Personally we don't mind the taste of seed hulls. We eat our sprouts hulls and all. If you throw away the hulls you throw away an excellent source of fibre.

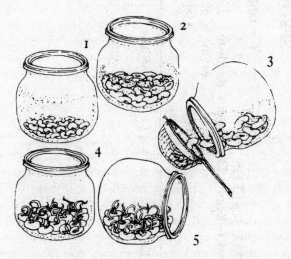

SPROUTING CHART

T = tablespoon, C = cupful

SMALL SEEDS
soak 6–8 hrs

	Dry amount to yield 1 litre/ 1¾ pints	Ready to eat in	Length of shoot (approx.)	Growing tips	Notes
Alfalfa	3–4 T	5–6 days	1½ in/ 3·5 cm	Delicious – one of our favourites! Taste particularly good after a day in sunlight.	Sometimes referred to as 'the father of all foods', alfalfa is one of the most nutritious and alkaline of all plants. Rich in organic vitamins and minerals, the roots of the mature plant penetrate the earth to a depth of 30–100 ft/9–30 m.

Fenugreek	½C	3–4 days	½ in/1 cm	Have quite a strong 'curry' taste. Best mixed with other sprouts.	Said to promote perspiration, therefore good for ridding the body of toxins.
Mustard no soaking needed	¼C	4–5 days	1 in/2.5 cm	Can be grown on damp paper towels for at least a week; the green tops are then cut off with scissors and used in salads.	Mustard is known as a counter-irritant and health tonic.
Radish no soaking needed	¼C	4–5 days	1 in/2.5 cm	Taste just like radishes! The red hot flavour is great for dressings, or mixed with other sprouts in salads.	Radish is particularly good for clearing mucus and healing mucous membranes.

SMALL SEEDS (*cont.*)

	soak 6–8 hrs	Dry amount to yield 1 litre/ 1¾ pints	Ready to eat in	Length of shoot (approx.)	Growing tips	Notes
Sesame		½C	1–2 days	Same length as seed	If grown for longer than about 48 hours sesame sprouts become very bitter.	Sprouting for just two days makes sesame more digestible and its nutrients (lots of calcium and vitamin E) more readily available.

LARGER SEEDS
soak 10–15 hrs

	Dry amount to yield 2 litres/ 3½ pints	Ready to eat in	Length of shoot (approx.)	Growing tips	Notes
Aduki beans	1½C	3–5 days	1–1½in/ 2.5–3.5 cm	Have a nutty 'legume' flavour.	Especially good for the kidneys.
Chick peas	2C	3–4 days	1 in/ 2.5 cm	May need to soak for about 18 hours to swell to their full size. The water should be renewed twice during this time.	Chick peas are a good source of protein and therefore helpful in body building for people who are underweight.
Lentils	1C	3–5 days	¼–1 in/ ½–2.5 cm	Try all different kinds of lentils – red, Chinese, green, brown. They are good eaten young or up to about 6 days old.	A staple food throughout the world – form a complete protein with rice and other grains.

LARGER SEEDS (cont.)
soak 10–15 hrs

	Dry amount to yield 2 litres/ 3½ pints	Ready to eat in	Length of shoot (approx.)	Growing tips	Notes
Mung beans	1C	3–5 days	½–2½ in/ 1–5 cm	Soak at least 15 hours. Keep in the dark for a sweet sprout. Put a weight (plastic bag filled with water and tied) on the beans to get long straight sprouts like the Chinese grow.	One of the most popular sprouts and one of the easiest to grow.

Soya beans	1C	3–5 days	1½ in/ 3.5 cm	Need to soak for up to 24 hours with frequent changes of water to prevent fermentation. An acquired taste. Be sure to remove any damaged beans which fail to germinate.	A good source of protein. Records show that as early as 3 000 BC soya beans were being used in the Orient.
Sunflower	4C	1–2 days	Same length as seed	Can be grown for their greens. When using sunflower seeds it is nice to soak them and sprout for just a day – that way they are more 'active' enzymatically. They bruise easily, however, and need to be handled with care.	An excellent food – supplies the entire body with valuable nutrients.

GRAINS

soak 12–15 hrs

	Dry amount to yield 1 litre/ 1¾ pints	Ready to eat in	Length of shoot	Growing tips	Notes
Wheat	2C	2–3 days	Same length as grain	A delicious sweet sprout with many uses including wheat grass and rejuvelac (the soak water). Large quantities are needed for breads.	An excellent source of the B vitamins. The soak water can be drunk straight, added to soups and vegetable juices, or fermented (rejuvelac, see page 200)
Rye	2C	2–3 days	ditto	Has a delicious distinctive flavour.	Good for the glandular system.

Barley	2C	2–3 days	ditto	As with most sprouts, barley becomes quite sweet when germinated.	Particularly good for people who are weak or underweight.
Oats soak 5–8 hrs only	2C	3–4 days	ditto	You need whole oats or 'oat groats'.	Like wheat, rye, and other grains, oats lose much of their mucus-forming activity when sprouted.
Millet soak 5–8 hrs only	2C	3–4 days	ditto	Must be unhulled millet, not couscous.	The only grain that is a complete protein and alkaline.

Mass production of sprouts

We find that the great demand for sprouts in our family rather outstrips the jar method. We need to grow them in much larger quantities and so we use seed trays, the kind gardeners use to grow seedlings in, with drainage holes in the bottom. We soak our seeds overnight as in the jar method, then rinse them well and spread them a few seeds deep on a double layer of plain white paper kitchen towels laid in the seedling trays. We then stand the seedling trays in a larger tray to catch the drips. We spray the seeds with a plant spray bottle and leave them in a warm place. They need to be checked morning and night and sprayed if they seem dry. If you wet them too much they rot. Larger seeds such as chick peas, lentils and mung beans need to be gently turned over twice a day to ensure that those underneath are not suffocated. Alfalfa seeds can be left alone – they grow into a thick green carpet after about five days. Then we put our trays of sprouts in a sunny place for a day or so to green up. When they are ready, we put them in a sieve, rinse them well and store them in airtight containers or sealed polythene bags until we want them. Sometimes we simply store them, tray and all, in the refrigerator, harvesting them

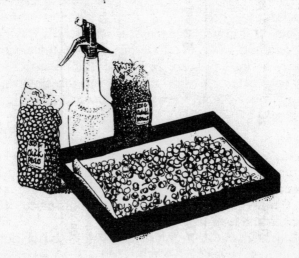

as we need them. To make the next batch, we rinse the trays well and begin again.

Easy and less easy sprouts and where to buy them
Some sprouts are more difficult to grow than others, but usually if seeds fail to germinate it is because they are too old and no longer viable. It is always worth buying top quality seeds because after removing any shrivelled or broken ones, and allowing for a percentage of germination failures, they work out better value than cheap ones. Try to avoid seeds treated with insecticide or fungicide mixtures, as sold by most gardening shops and nurseries. Health food shops are usually your best bet. It is fun to experiment with growing all kinds of sprouts from radish seeds to soya beans. But don't be tempted to grow greens from seed potatoes or tomato seeds – they are poisonous, and belong to the deadly nightshade family! Steer clear of kidney bean sprouts too – both they and the raw beans they grow from are poisonous.

Some of the easiest sprouts to grow are: *alfalfa seeds*, *aduki beans*, *mung beans*, *lentils*, *fenugreek seeds*, *radish seeds*, *chickpeas* and *wheat*.

Once you have tried some of these you might like to graduate to: *sunflower seeds*, *pumpkin seeds*, *sesame seeds*, *buckwheat*, *chia*, *flax*, *mint*, *red clover seeds*, most beans and peas (*soya*, for example), *peanuts*, *almonds*, *tricitale*, *rye*, *oats*, *sweet corn* and *millet*. We have put these in the advanced category because they can be difficult to find or to sprout. They must have unbroken hulls, and the nuts, almonds and peanuts must be really fresh and undamaged. Good luck!

Growing seedlings in soil
This is a way of growing seeds beyond the sprout stage and into tiny succulent salad plants, and it takes approximately two weeks. By then you will have seedlings about 6 in/15 cm high. The seeds most suited to this treatment are *wheat*, *buckwheat* and *sunflower*.

Fill a seed tray with moist (not soggy) soil or potting

compost. Waterlogged soil only encourages fungus to grow on the seeds. Cotton wool, paper towels or even cotton towels can be used instead of soil – less messy and more convenient if you do not have a garden – but soil is preferable because it contains useful minerals which you will get the benefit of. Soak your seeds for the appropriate length of time, then let them sprout in a jar until little white roots begin to show. Now sprinkle them, one layer thick, on the soil and cover with another tray or with a sheet of black plastic. Leave for three days in a warm place.

On the fourth day put the seedlings in the light, but not in direct sunlight. Spray them with water and continue growing them for up to two weeks or until they reach a height of about 6 or 8 in/15 or 20 cm, watering when necessary. Wheat grass is sweetest when the stems begin to branch.

To harvest, cut the seedlings as close to the seed hulls as possible. If you continue watering the stubble you may get a second batch of greens. Some seedlings, buckwheat, lettuce and sunflower greens for example, are good to eat whole; just rinse them to remove any earth. Wheat grass is better chopped finely and sprinkled over salads or juiced along with other vegetables.

The yoghurt story

Yoghurt is most important for its action on the flora or micro-organisms which colonise the human intestines. Once in the intestines the lactic acid bacteria which yoghurt contains get to work synthesising valuable B vitamins, and in doing so they make conditions in the colon sufficiently acid to discourage the growth of pathogenic and putrefactive bacteria. In fact, laboratory studies show that many pathogens, such as those which cause typhoid fever, dysentery and diphtheria, lose their virulence when placed in yoghurt and are even killed in yoghurt whey. This is one of the reasons why yoghurt is very good for curing gastro-intestinal disorders. It is also helpful in restoring the digestive tract to its normal condition after the use of antibiotics. Useful though they may be in real emergencies, antibiotics destroy all intestinal bacteria, including the friendly ones, and many of the B vitamins present in the gut as well.

Yoghurt is much more easily digested than milk. One reason for this is that the milk protein in it has been partially broken down by bacteria. Perhaps even more important is the breakdown of lactose (milk sugar) to lactic acid which occurs when milk is made into yoghurt. This is of significance because many people (whether they know it or not) are not able to digest this sugar. In adulthood they lose the ability to produce the enzyme lactase so that they can no longer break down lactose. This results in lactose intolerance. Undigested lactose stays in the intestines and attracts water. In extreme cases it can cause bloating, excessive flatulence, abdominal pains and diarrhoea. Nevertheless many people who cannot tolerate milk can eat yoghurt without any problem. Another advantage yoghurt has over milk is that the calcium and phosphorus in it are much more available for absorption.

The best yoghurt is made from sheep's or goat's milk. Cow's milk is harder to digest and more mucus-forming. Goat's and sheep's milk and yoghurt are sold in health food shops, plain natural cow's milk yoghurt in most supermarkets.

How to make your own yoghurt

It is really worthwhile to make your own yoghurt because, provided you can get good fresh goat's or sheep's milk to make it from, it need not be heated above body temperature, which means that none of the health-giving enzymes in it need be destroyed. Also, home-made yoghurt tastes so much better than the bought kind. One reason is that manufactured yoghurt is not as fresh as it could be and sometimes contains stabilisers and preservatives to prevent it going off too quickly. The result is a slightly sour taste which some people find off-putting. The natural home-made kind is actually sweet tasting. With a little practice you can quickly become an expert at making it.

Yoghurt-making is really a lot easier than most people

think. You do not need fancy yoghurt makers, thermometers or sterilising fluids. All you need are some fresh milk, a container, a warm place and a 'starter'.

Milk Goat's and sheep's milk are best, and there is nothing wrong with buying a large quantity and freezing it. Soya milk can also be used. If you want to make cow's milk yoghurt you can even use low fat skim milk powder, marginally preferable to whole cow's milk and very simple to use as it does not need to be boiled.

Containers Use whatever containers you happen to have in your cupboards, either an earthenware pot or casserole, a heat-resistant wide-mouthed glass jar, a wide-mouthed thermos flask or a stainless steel saucepan. Always go for inert materials – no aluminium or flaky lacquered dishes. Preferably your container should have a lid, but cling film will do.

A warm place There are many ways of getting round this one. An Aga stove is ideal. You can stand your container directly on it or on a wire cooling tray if it is too hot. An airing cupboard or an oven heated to 120°F and then switched off are good alternatives. If you choose a radiator, or the warm area at the top back of the fridge, wrap your container in a blanket or towel for insulation. You can also use a polystyrene bucket or picnic hamper with a lid to make an 'incubator'. Or you could make a simple 'hay box' using a couple of cardboard boxes: you fill one with hay, or torn up newspaper or some other insulating material, put the yoghurt container in it, and fit the other box over the top as a lid. The ideal temperature to be maintained is 90–105°F.

Starters There are two kinds of starter, plain yoghurt or a powdered culture. If you use yoghurt starter it does not matter too much if it is cow's rather than goat's or sheep's, but it should be plain, natural yoghurt with nothing added. Read the labels! Some things advertised as yoghurt in supermarkets in fact contain no lacto-bacteria at all. Don't buy fruit yoghurt – it doesn't work! Once you have made your first batch of yoghurt you can use your own yoghurt as a starter indefinitely. In fact, you will find your yoghurt getting tastier each time you make it. If the flavour starts to become sour, use a fresh starter. Powdered yoghurt starter can be bought in some health food shops.

Yoghurt – step by step

Once it has been heated to body temperature and the starter added milk turns into yoghurt in 10 hours or less. This is what you do.

● Heat 2 pints (1 litre) of milk to just below boiling point, or until small bubbles start appearing around the edges of the saucepan. You can buy a round glass or china disc that sits in the bottom of the pan and begins to rattle as boiling point approaches. If you are using fresh goat's or sheep's milk from a good supplier, you can skip this step and just warm the milk to body temperature. If you are using milk powder blend it thoroughly with pure water at blood heat. The more powder you use the thicker your yoghurt will be.

● Leave the milk to cool down sufficiently for you to be able to put a finger into it without discomfort. It should feel neither hot nor cold – but about blood heat.

● Rinse your container with boiling water. This sterilises it, which is important because you don't want any foreign bacteria in your yoghurt. It also warms it and helps keep the milk at a constant temperature while the yoghurt is incubating.

● Pour the milk into the container and add your starter, a generous tablespoon of yoghurt per pint or half litre of milk or the quantity of powdered starter stated on the packet.

● Stir the starter in well. This is important because it distributes the bacteria evenly, otherwise you end up with a lump of yoghurt swimming in a sea of milk! Be sure that whatever you use to stir the mixture has been rinsed in hot water too.

● Place the lid on the container, or cover with cling film – yoghurt bacteria require oxygenless working conditions. Put the container in a warm place and leave for about 5 or 6 hours. The faster the yoghurt ferments, the sweeter it will be. If it hasn't curdled in this time, leave it longer. At all events it should not take longer than 10 hours. Experiment with the temperature of your warm place; a slightly higher temperature will speed things up.

Goat's and sheep's milk yoghurts tend to be thinner than cow's milk. If you get a rather watery yoghurt first time, don't worry. It is still delicious and in any case it tends to get thicker every time you culture a new batch. Yoghurt keeps in the refrigerator for up to a week.

Ferments and cheeses

Fermented foods are not only delicious, but valuable for health as well. Rich in enzymes and lacto-bacillus bacteria, the proteins they contain are predigested and therefore very easy for the body to assimilate. The lactic acid in them destroys harmful intestinal bacteria and friendly bacteria boost vitamin production in the intestine.

Rejuvelac This is simply the fermented soak water from wheat sprouts. It can be drunk straight to clean the system and improve the condition of the intestinal flora, or it can be used to make seed and nut cheeses and other ferments. Rejuvelac contains eight B vitamins as well as vitamins E and K. Because its protein and carbohydrate contents are in the form of amino acids and simple sugars its nutrients are readily assimilated.

One cup of wheat grains will yield approximately nine cups of rejuvelac. For the first soaking use one cup of wheat grains and three cups of pure water and pour off the rejuvelac after 48 hours. Using the same wheat grains repeat the operation three more times, each time using two cups of pure water and soaking for only 24 hours before pouring off. Each yield of rejuvelac should be refrigerated if not used immediately. At the end of this time the wheat grains can be eaten, made into Essene bread (recipe page 264), or sown to grow wheat grass. Rejuvelac should taste quite sweet, not sour or tart. If it tastes unpleasant it has probably over-fermented. It will keep in the refrigerator for up to five days.

Seed/nut cheeses These are truly delicious and unusual concoctions made of ground seeds and nuts, rejuvelac and various seasonings. The ingredients are mixed together and left to ferment for about 12 hours, and the result is a subtle tangy soft cheese.

One of our favourite seed/nut cheeses is *almond cheese*. The ingredients are 1 cup of blanched almonds, 1 cup of sunflower seeds, 1 cup of rejuvelac, seasoned with a little vegetable bouillon, a tablespoon of tahini (optional) and a few finely chopped spring onions or chives, parsley and other herbs.

Grind the almonds and sunflower seeds to a fine meal, add the rejuvelac and combine well. Stir in the chopped herbs, bouillon and tahini. Place in a bowl covered with a tea towel and leave to ferment for about 12 hours in a warm place.

This basic recipe has infinite variations, depending on the nuts, seeds and seasonings that appeal to your taste. Cashew nuts and sunflower seeds make an excellent combination and so do walnuts and sunflower seeds. Some other winning flavourings are sage and onion or garlic and fines herbes.

FOUR

RAW ENERGY
RECIPES

BON APPÉTIT!

This is where we unveil the full colour and zing of our favourite raw energy recipes. Many people say to us: 'Your diet must be so boring! If you don't eat meat, fish, poultry or cheese, or very rarely, and if you don't eat bread, cakes, biscuits or pasta, and if you don't eat any tinned or convenience foods, what on earth do you eat?' Well, see for yourself.

Raw Energy offers a whole new world of culinary excitement. It is not merely a lesser known branch of the art, to be added to or omitted from the serious business of food preparation, but a delightful alternative, with its own techniques and showmanship. Raw food cuisine has its own ingredients – fresh fruits and vegetables of every description, nuts, seeds, sprouted ingredients, ferments – just as conventional cuisine has things like flour, milk, eggs, butter, sugar, brandy, meat and so on. Like the conventional chef, the raw food chef can prepare an unending variety of dishes from these basic ingredients.

The immediate appeal of raw food dishes is their variety of texture – from crunchy crudités to creamy dips – and their gorgeous colours, the pure colours found only in nature. Add to this the myriad aromas of fresh herbs and spices and unusual condiments and one begins to sense the wonderful possibilities of raw cuisine.

Eating raw food does not mean sacrificing a three or four course meal for a plate of lettuce. If you are entertaining you can prepare a raw banquet every bit as magnificent as the Lord Mayor of London gives every year. Raw food is not the dieter's curse but an exotic and innovative way of eating. And you do not need all the time in the world to make a raw meal. You can toss together a delicious Dish Salad (see page 219), a full meal in itself, in a matter of minutes.

Even more glorious is the fact that in raw cuisine there can be no 'failures'. If something is not the consistency you wanted it, change it into something else. A thin mousse gracefully turns into a dessert soup if you call it by another name!

Measurements and quantities

The quantities given in our recipes serve about four people, but they can easily be adjusted to suit your requirements. Quantities are never critical in raw cuisine because there are no recipes where success or failure hangs by a whisker. We give approximate measurements such as cupfuls (abbreviated to C – an ordinary cup holds about 8 fluid oz), handfuls, tablespoons (T), teaspoons (t), pinches, and so on.

Each time you make a recipe it will be slightly different, which is the whole fun of cooking and eating. Sticking to cookery book rules takes all the spontaneity out of things. Embroider our recipes to suit your taste and imagination. You will quickly grasp the basic principles of texture, colour, nutritional value, taste combinations and presentation. Bon appétit!

HORS D'OEUVRES AND MINI SALADS

Appetisers should be both beautiful and delicious. When you sit down to eat your first mouthfuls they should prepare your body for the meal to come by stimulating digestive secretions through sight and taste. They also serve as a punctuation for the meal to follow, reminding you to relax and enjoy the pleasure of eating. People are often so keyed-up by stress and other problems that they forget to unwind and hardly taste their food. They gulp it down unsmelled, untasted and half chewed. Of course they get indigestion! And because they do not give their digestive machinery a chance, they get little nutritional value out of their food. So be creative preparing your hors d'oeuvres and make each one a feast for the eyes as well as the taste buds!

Stuffed Things

GREEN CREPES

These are stuffed lettuce leaves, but the leaves need to be large and flexible so that they roll without splitting. For the stuffing we use finely chopped vegetables in a creamy dressing. A combination we particularly like is:

alfalfa sprouts, avocado, tomato, red and green peppers, spring onions, finely grated carrot or beetroot, cucumber

Finely chop or grate the vegetables and mix them together with the dressing of your choice – a thick creamy mayonnaise (see page 237) is very good. Put spoonfuls of the mixture onto the lettuce leaves, roll them up and spear with a cocktail stick to hold them in place.

STUFFED TOMATOES WITH FRESH BASIL

The best tomatoes are the home-grown sort, picked when they are red and ripe and full of flavour. The commercial sort are picked while they are still green and hard because they travel better, and are then treated with ethylene gas which turns the skin red but leaves the inside unripe and often flavourless.

4 large tomatoes, ¹/₂C tahini mayonnaise (see page 238), 2 spring onions, 8 leaves fresh basil, ¹/₂t French mustard, ¹/₂ clove garlic (pressed)

Slice the tops off the tomatoes and keep them. Scoop out the seeds and pulp and chop with the onions and fresh basil. Mix with the tahini mayonnaise, mustard and garlic. Spoon the mixture into the tomato shells. Cut the lids in half and stick them into the top of the filling to make butterfly wings. Serve each tomato on a lettuce leaf. Guacamole (see Iguana salad on page 223) without the extra tomatoes also makes a delicious stuffing. So do most of the dips on pages 244–6.

PRUNE STUFFED CELERY

*4 sticks celery, 8 prunes (soaked overnight in a bowl of water), ½C
seed cheese (see page 200) or cottages cheese or cream cheese, 1
orange, a little nutmeg*

Trim away the base of the celery stalks, but leave their tops
on. Stone and chop the prunes, and mix them with the
cheese, nutmeg, and the juice of half the orange. Fill the
celery hollows with this mixture. Peel the other half of
the orange, slice it and use it to decorate each celery stick.
Celery sticks are also good filled with stiff tahini mayonnaise
seasoned with vegetable bouillon and sprinkled with paprika.

SEED CHEESE STUFFED PEPPERS

This recipe looks most attractive with red peppers, but you
can use green. Green peppers turn red as they ripen and
become sweeter in taste. The idea is to stuff them with a firm
cheese and herb mixture so that they can be sliced crossways
and served as red- or green-edged medallions. You can also
serve them in halves as a main course, because they are quite
filling.

*2 red peppers, 2C seed and nut cheese (either cashew and sunflower
or almond and sunflower – see page 201), chives, fresh parsley,
lovage or marjoram, 1t vegetable bouillon*

Remove the stalks and seeds from the peppers by cutting
around the stalks with a sharp knife. Be sure to scoop out all
the seeds, and the white pith as well. Finely chop the chives
and other herbs, mix with the seed cheese, not forgetting to
add the bouillon powder. Fill the peppers with the mixture
and pack down firmly. Wrap each pepper in cling film and
chill for about an hour, by which time they should be firm
enough to slice crossways into medallions. Serve on a bed of
curly endive garnished with sprigs of parsley.

If you want to make stuffed pepper halves simply slice the
peppers in half lengthways to begin with, remove the seeds

and pith, and fill with the cheese mixture. Serve sprinkled with a little paprika and on a lettuce leaf.

STUFFED FAIRY CUPS

Button mushrooms are simply the immature predecessors of the large flat-topped brown-gilled mushrooms which come into the shops in autumn. Ideally they should be white, unbruised and perfect, straight from the mushroom farm. You can of course buy a mushroom kit and grow your own in a large bucket successfully and cheaply.

8–12 large button mushrooms, 1/4C alfalfa sprouts or finely grated hard cheese, 1/4C almonds (ground), 3T yoghurt, a squeeze of lemon juice, 1t honey, 1t dill seeds (roasted and crushed), fresh parsley and mint, vegetable bouillon powder

Remove the stalks of the mushrooms, trim and keep. Grind the almonds as finely as possible and mix with the yoghurt, lemon juice and honey. Add the ground dill seeds and a little chopped parsley. Chop the mushroom stems finely and mix in. Spoon this mixture into the mushroom cups and sprinkle with alfalfa sprouts or finely grated hard cheese. Serve twos or threes on little dishes garnished with mint sprigs.

Vegetable 'Flowers'

Here are a few decorative ideas which can raise any hors d'oeuvre, or any main dish for that matter, several notches up the 'eye appeal' scale.

CARROT OR COURGETTE ROSETTES

Wash your carrots or courgettes, then with a very sharp knife cut about six grooves, regularly spaced, along their length and remove the long thin slivers of flesh. Slice them crossways into little flowers.

TOMATO LILIES

There are two ways of making these; the first is better for smaller tomatoes, and the second for larger, slightly firm ones.

For small tomatoes, place them stem end down on a cutting board, and cut almost all the way through them with four crossing cuts so that you end up with eight petals. Gently pry the petals open.

For larger tomatoes you make small zig-zag cuts around their equators cutting right through to the centre with each cut. Your knife needs to be very sharp to do this. The smaller the angle between your zig-zags the more petals you will have and the more attractive your lilies will look. When you have cut all the way around, separate the two halves.

RADISH ROSES

Again, there are two ways to tackle these. If your radishes are long, simply cut a criss-cross pattern of four cuts in the bottom end (the end opposite the stalk), cutting to about half way down each radish. Then place the radishes in a glass of cold water in the fridge for the petals to open up. If you have larger, chubbier radishes you can cut zig-zags around the middle as for Tomato Lilies, then separate the halves.

BEET GARDENIAS

Small young beets are best for this, but you can also use small turnips. Trim the root and place on a board, root end up. Using a sharp knife make a series of vertical cuts in one direction, then another series of cuts at 90° to the first. Take care to keep your cuts close together and deep, but not so deep that you cut through to the bottom. Place the beets in the fridge in a bowl of cold water for about half an hour for the petals to open.

Crudités

One of our favourite hors d'oeuvres or snacks is a platter of crudités, crunchy raw vegetables and fruit sliced or chopped so that you can pick them up and eat them with your fingers and garnished with lemon slices, sprigs of watercress, parsley, mint or sprouts. They can be eaten dipped into a sauce (see dips page 244) or simply sprinkled with a light dressing and a few toasted fennel or caraway seeds. The important thing is how you prepare them.

STICKS AND MATCHSTICKS

These are very tempting. Carrots, turnips, courgettes, cucumbers, celery and pineapple are naturals for making sticks. To make matchsticks just keep chopping lengthwise until you get sticks as thin as matches. Green or red peppers also make good matchsticks. To keep sticks fresh, put them in a bowl of cold water with a squeeze of lemon juice in the refrigerator but don't keep them in water for longer than half an hour.

SLICES

Some vegetables are particularly nice sliced diagonally. This makes larger pieces for better 'dunking'. Try diagonal slices of cucumber, carrot and white radish. Very thin slices of small beetroots, Jerusalem artichokes, kohl rabi and turnips are also nice. Large apples sliced crossways can be used as 'bread' for open sandwiches. Sweet peppers cut crossways make attractive rings. Try cutting 'serviette rings' of pepper and placing them round bundles of carrot or celery sticks.

WHOLE VEGETABLES

Button mushrooms with their stalks on, whole baby carrots, the sweet centre shoots from a head of celery, whole young green beans which have been topped and tailed, cauliflower florets, radishes, young spring onions – all of these make delicious crudités. All you do is trim and rinse them, and dry on a paper towel.

WEDGES

Orange and tangerine segments make natural wedges for a platter of crudités. But so do tomatoes, chicory, Webb's Wonder lettuce, apples and pears if you cut them right.

SALAD KEBABS

Skewered rows of different vegetable and fruit chunks make a very unusual starter. Make *salad kebabs* by skewering a row of different vegetable and fruit chunks. Choose from:

button mushrooms, pitted olives, cherry tomatoes, pepper chunks, cucumber or courgette chunks, fresh garden peas, tofu chunks, soaked dried fruit (prunes, apricots)

Serve your skewers with a spicy dressing on a bed of shredded lettuce. Mmmmm!

ON COCKTAIL STICKS

Thread small chunks of fruit and/or vegetables on cocktail sticks and stand them up, porcupine fashion, in half a grapefruit, half a small cabbage, or the stumpy centre of a cauliflower as a party piece. Skewer each stick with several different items, such as:

cubes of cheese, apples, cucumber, celery, pineapple, small shallots, small radishes, pitted olives, grapes, apricot halves, raisins, chopped dates

Fruit Hors d'oeuvres

AVOCADO IDEAS

Avocado pears make attractive and delicious appetisers. There is so much more you can do with them than the familiar 'avocado vinaigrette'. Many people do not realise that avocados are a fruit, although they contain very little fructose, and that they go particularly well with other fruits. In fact the Brazilians eat them with cream and sugar! Even the shells have their uses; we often fill them with Iguana Salad (see page 223) as appetisers.

CRUNCHY APPLE AVOCADOS

2 ripe avocados, 2 dessert apples (preferably red) finely diced, 1 handful walnuts or pecans, 1 handful raisins, 1 stick celery very finely chopped, juice of ½ lemon, 3T plain egg mayonnaise, cinnamon

Cut the avocados in half and scoop out the flesh. Keep the shells. Mash the flesh with lemon juice to prevent it going brown. Roughly chop the walnuts. Chop the apples and celery finely. Mix apples, celery, walnuts, raisins and avocado together, and add the mayonnaise and a pinch of cinnamon. Fill the avocado shells with the mixture and garnish with a walnut and an apple slice.

AVOCADO OR PEAR APPETISER

Either avocado halves or pear halves can be used for this. If you are using pears make sure they are ripe so that you can scoop an avocado-sized hole in each half.

2 avocados or pears, 1T lemon juice, ½C yoghurt, 2t honey, ½C fresh fruit finely chopped (pineapple or black cherries are particularly good), lemon juice, ground ginger

Halve the avocados or pears and take out the stones or cores. Brush the cut surfaces with lemon juice. Mix the yoghurt, honey, chopped fruit and ginger together and spoon into the hollows. Serve each half on a leaf of lettuce.

ORANGE AND WATERCRESS

Simply peel and finely slice one orange per person (if you have time, chill the oranges in the fridge for half an hour before peeling) and garnish with a few watercress sprigs.

Sprinkle with a light dressing (see pages 237–244) just before serving.

CYPRIAN OLIVES

Olives are often terribly salty. We pour off the pickling water they come in and store them in a jar with vegetable or olive oil until we need them. This not only makes them taste less salty, but gives extra flavour to the oil which can be used in salad dressings.

Put a handful of shiny black olives on each plate and sprinkle them with finely grated lemon rind. The rind of one lemon is enough for four servings. Then sprinkle with dried oregano flowers, if you can get them.

Cocktails – as Hors d'oeuvres

It makes a refreshing change to start a meal with a liquid appetiser made from fresh fruits and vegetables. Any of the raw juices on pages 277–82 can be served as cocktails to start a meal. The process of juicing or blending releases lots of enzymes and nutrients, so you should serve juices and cocktails as soon as you make them or much of their high energy value will be lost by oxidation. Adding fresh herbs encourages the stomach to produce its digestive juices. Here are two of our favourite cocktails.

ORANGE PINEAPPLE COCKTAIL

Pineapple is an excellent aid to digestion because it contains the enzyme bromelain which encourages the secretion of hydrochloric acid in the stomach and helps to digest proteins. In Hawaii pineapple is often eaten at the end of a meal as a digestive.

1 small pineapple, 1 or 2C orange juice, several ice cubes, ground ginger, fresh mint, lemon slices

Peel the pineapple, cut it into chunks and process in the blender or food processor with the orange juice. Add the ice cubes, ginger and mint and blend again. Serve immediately in tall or stemmed glasses with a slice of lemon hooked over the rim.

MIXED ENERGY COCKTAIL

2 carrots, 2 tomatoes, 2 sticks celery, fresh parsley, basil and marjoram, 1 lemon, 1t vegetable bouillon powder, several ice cubes, and enough water to give the desired consistency

Peel the tomatoes. Roughly chop the carrots, celery and tomatoes and put them in the blender. Add the lemon juice, herbs, bouillon powder, water and ice cubes and blend. Serve in glasses with a sprig of fresh parsley, a slim swizzle stick of cucumber and a dash of tabasco or Worcestershire sauce if desired.

SOUPS

Soups make an excellent first or even second course. A blender/food processor/juicer is invaluable for soup making. It saves time but it also allows you to achieve the smooth consistency everyone associates with soup and a thorough blending of all your flavours.

Many raw soups have a vegetable juice base. Even people who wouldn't dream of drinking a glass of carrot juice can be seduced by the subtle flavours of 'Carrot Chowder'. Whereas many cooked soups have a rather washed-out appearance, raw soups have brilliant colours that will delight and entice you and your family. We make fruit as well as vegetable soups, but both are a tasty and nutrient-packed addition to a main meal. If you prefer your soups hot rather than cold, simply heat them gently to 90°F and serve. This way you do not destroy the enzymes.

Vegetable Soups

CARROT CHOWDER

3C carrot juice (you'll need about 12 big carrots for this), 1C nuts (almonds, hazels or pecans), 1C goat's milk yoghurt, 2 egg yolks, 1T olive oil, juice of ½ lemon, small clove garlic, ½ green pepper, 2 spring onions, chopped parsley, 2t vegetable bouillon powder, ice cubes

Grind the nuts finely and blend them with the yoghurt, egg yolks, pressed garlic, lemon juice, olive oil and seasoning. Juice the carrots into a jar with ice cubes in it, then slowly add the juice and ice to the yoghurt mixture, stirring well. Serve sprinkled with a mixture of finely chopped green pepper, spring onion and parsley.

FLAMINGO SOUP

2 medium sized beetroots, 10 carrots, 1 small head celery, 4–6 tomatoes, 2 handfuls almonds, 1T fresh thyme, 1T fresh basil, vegetable bouillon to taste, ice, juice of 1 lemon, 6T yoghurt, chives

Juice the beets, carrots and celery and put in an airtight jar with some ice and the lemon juice. Blend the tomatoes, almonds, thyme, basil and bouillon powder. Combine the two mixtures and serve in bowls with a spoonful of yoghurt and a sprinkling of chopped chives.

FRESH GREEN SOUP

2 avocados peeled and stoned, 3C apple juice, juice of 1/2 lemon, 1 courgette, a handful of alfalfa and mung sprouts, 1 stick celery, parsley, 2t tamari, 1t vegetable bouillon powder, ground ginger, sliced mushrooms or flaked almonds

Combine the avocados, apple juice, lemon juice, parsley, tamari, vegetable bouillon and a pinch of ginger in the blender. Grate the courgette and finely dice the celery and mix them with the sprouts. Now pour on the avocado sauce. Serve sprinkled with sliced mushroom or flakes of almond.

POT LUCK SOUP

This is made with whatever vegetables you happen to have available, hence its name. Here is one possible combination:

1/4C of each of the following: green beans (finely chopped), carrots (grated), celery (finely diced), fresh garden peas or cauliflower florets or broccoli tips (finely chopped), 1 handful of sprouts, 1/2 mild onion or 2 spring onions (finely chopped)

Mix all these together in a bowl and cover it with cling film while you prepare the 'liquid' ingredients of the soup. These are:

4 tomatoes, 2C fermented seed and nut dressing (see page 240) or 2C vegetable juice, 1/2 avocado, 1/2 clove garlic, 1T tamari, a little vegetable bouillon

Put all these ingredients in the blender, then pour the liquid over the chopped or grated ingredients. Serve in individual bowls.

GASPACHO

This soup is a combination of smooth and crunchy. It is particularly good served with 'croutons' – roasted soya nuts, wheat and barley (see page 286).

2 small cucumbers, 1T minced onion, 3 tomatoes, 1 red pepper, 3 egg yolks, 3T vinegar, 3T olive oil, 1 clove garlic, ½C tomato juice, 2 spring onions, fresh parsley and basil, 2t vegetable bouillon, 1t honey, dash of red wine (optional)

Purée the onion, tomatoes, red pepper and one of the cucumbers in the blender, then add the egg yolks, olive oil, vinegar, garlic, tomato juice, honey, seasoning and wine. Finely chop the spring onions, the other cucumber and the fresh herbs. Add these just before you serve the soup. Put the 'croutons' in a separate dish so that people can help themselves.

Fruit Soups

BLACK CHERRY SOUP

2C sweet black cherries, juice of 3 oranges, 2T honey, 2C water, grated coconut, fresh mint sprigs

Pit the cherries and blend with the orange juice, honey and water. Serve chilled, decorated with mint sprigs and some grated coconut.

STRAWBERRY SURPRISE SOUP

2 small punnets strawberries, 2½C milk, ½C cashew nuts, 1t vegetable bouillon, ½t nutmeg, ½t ginger, fresh mint

Put the nuts, milk, nutmeg, ginger and bouillon in the blender and mix until smooth. Blend the strawberries to a cream (keeping aside a few for decoration). Add strawberry mixture to the milk mixture and stir in well. Serve in individual bowls garnished with fresh chopped mint pieces and slices of strawberry.

SALADS AS MAIN DISHES

The salads in this section are a far cry from the one-limp-lettuce - leaf - slice - of - cucumber - tomato - and - dollop - of-salad-cream monstrosity which lurks on restaurant mènus under the guise of 'salad'. The Oxford English Dictionary quotes a discerning Briton who in 1846 said: 'The salad is the glory of every French dinner and the disgrace of most in England.' It is high time the salad became the glory and delight of *every* dinner.

As the central dish of all Raw Energy main meals, salads need to be varied, substantial and most emphatically delicious. The key to a good salad is fresh flavourful ingredients. Which to combine together is obviously important, but also how to cut them up, how to dress them and how to present them. There are dozens of little tricks which help you to make mouth-watering, extraordinary salads from very ordinary ingredients, and many ways of combining tastes you never thought went together in the same dish.

Delicious Dish Salads

These are our favourites. They can be a meal in themselves and are quick and easy to prepare. They are made in individual dishes, one for each person. You çan create a salad to suit the specifications of each member of the family just as easily as making one large salad! You simply put little piles of vegetables, fruits and sprouts in each dish and pour on the dressing or dip of your (or their) choice. The important thing is how you cut your vegetables and fruits – this applies to all salads.

● We often use greens such as lettuce or Chinese leaves finely shredded as the base of a salad; the dressing and juices run down into it so that it is especially tasty when you get to it.

● Carrots, cucumber and celery are nice chopped into

sticks and used to scoop up creamy dips. Cut the celery twice lengthwise into three strips, then cut once crosswise and you have six sticks of even size. Leave the leaves on – they are attractive and tasty. Carrots take a bit more care. Cut them in half lengthwise, then lay a half flat side down and cut it in half again, then cut each of the quarters in half lengthwise, so you have eight sticks. As with all kinds of vegetable sticks, the thinner they are, the better they taste. If you take a little bundle of sticks and chop them across, you get neatly diced vegetables.

• Carrots, like most root vegetables, are also good grated. Try the coarse and fine attachments of your food processor for variation. Also try grating beetroots, turnips, parsnips, swedes, celeriac, Jerusalem artichokes, kohl rabi, radishes, horseradish, Brussels sprouts, cabbage (red and white), cauliflower and broccoli stems, fennel, courgettes and mild onions.

• Courgettes, aubergines and some cucumbers have a bitter flavour which can be got rid of by slicing them and soaking them in a little cider vinegar. Leave for about ten minutes so that the bitter juices are drawn out. Then wash the slices thoroughly under a cold tap to rinse away the vinegar, and pat dry. Inevitably by this method, or by the salt sprinkling method, one loses some vitamins and minerals.

• Tomatoes are nice chopped or sliced in salads, but don't forget to remove the hard green stem area. Their skins can be removed, for special dishes, by dropping them in boiling water for half a minute, or spitting them on a fork and turning them quickly over a naked flame. The skins split and shrink with the heat and can easily be pulled off.

• Half an avocado chopped into a salad gives a pleasant smoothness which contrasts with the crunchiness of most of the other vegetables. To dice an avocado quickly, cut it in half and remove the stone, then hold one half in your hand and make several cuts first lengthwise and then crosswise. Scoop the chopped flesh into your salad with a spoon.

• Spring onions, chives and other fresh herbs are nice chopped very finely. With spring onions cut them two or three times lengthwise first, and then across.

● Use fruit in dish salads – sliced orange, apple or peach – to add flavour and colour. A handful of seedless raisins is good too; try chilling them in the fridge first – they become quite chewy.

● Sprouts are a must – they really do make a dish salad. Alfalfa sprouts are good as a base and chick pea, mung, lentil, aduki and fenugreek sprouts combined together make a dish salad on their own.

Have a look at Salad Sprinkles at the end of this chapter and Vegetable 'Flowers' on page 209 for more ideas. Don't be limited by what you have seen or heard. Experiment with chopping, grating and slicing different vegetables. Usually the more finely cut a vegetable is, the more flavour it has and the more delicious the salad. A splendid dressing of course is the pièce de résistance of a good salad! There are plenty of mouthwatering suggestions on pages 237–46.

Apart from being meals themselves, dish salads are great for dieters and anyone who likes to 'pick' between meals. Put all those things you like to pick into a dish salad and eat them all together. That way you will know exactly how much you are eating, and you will feel satisfied, and less tempted to pick, afterwards.

Mixed Salads

COUNTRY GARDEN SALAD

This is especially appealing because of the different textures and flavours of the vegetables. Raw young fresh peas are surprisingly sweet and tasty. At one time peas were a luxury item, much as oysters or fine chocolates are today. In the seventeenth century fashionable people who over-indulged ate peas by the plateful before retiring to bed!

1 small Webb's Wonder (iceberg) lettuce, 2 carrots (diced), 1C finely shredded red cabbage, 1C fresh garden peas, 1 finely chopped shallot (optional), 2 tomatoes (chopped), 1 courgette or small cucumber (sliced paper thin), a handful of alfalfa sprouts

Wash the lettuce first and put it in a polythene bag in the fridge to crisp up while you prepare the other vegetables. Tear the lettuce into small pieces and mix with the rest of the salad items. Toss with a light dressing such as French or Citrus (see pages 239–40).

SUMMER SYMPHONY

This salad is an exciting play of colours and shapes – the more variety the better.

1 lettuce (cos is good), 1 C small cauliflower florets, 2 celery stalks (finely chopped), 2 carrots (finely grated or cut into matchsticks), 6 cherry tomatoes, 4 radishes (sliced), 1 green pepper (cut into thin strips), watercress, fresh sweet corn or alfalfa sprouts to garnish

Place the lettuce leaves, torn into bite size pieces or shredded, into a bowl – a clear glass bowl is nice for this one so that all the beautiful colours show through it. Prepare the vegetables and arrange in layers in the bowl, keeping the watercress for decoration. Dress with a thinned mayonnaise dressing, perhaps blended with a tomato or two, and top with sweet corn or alfalfa sprouts, and sprigs of watercress.

ITALIAN SALAD

If you can get hold of the ingredients, this salad is something special.

1 Italian red lettuce (radicchio), 1 small cos lettuce or 2 chicories (finely shredded), 1 red and 1 yellow sweet pepper (cut into rings or diced), 1 or 2 large Italian tomatoes (peeled and diced), 4 radishes (cut into segments), 1 red onion (cut into thin rings), button mushrooms (thinly sliced), fresh basil, fennel seeds or black pepper

Shred the lettuces and chicories and put them into a large bowl. Make a nest in the centre and put the other vegetables into it, sprinkling the onion and mushroom slices in last. Toss with a spicy Italian dressing with lots of fresh basil and sprinkle with toasted fennel seeds or freshly ground black pepper.

IGUANA SALAD

This is one of many variations of Guacamole, a favourite Mexican dish based on avocados. Iguana Salad can be used as a dip, scooped into Green Crêpes, or eaten straight . . . any way it is a real treat. To ripen hard avocados, put them in a brown paper bag in a warm spot for a couple of days.

2 or 3 really ripe avocados, 2 juicy tomatoes, ½ red pepper, 1 small onion (finely grated), juice of 1 lemon or lime, small clove garlic, tabasco sauce and/or 1 fresh chilli (very finely chopped), olive oil, pepper and vegetable bouillon to season

Chop the tomatoes and red pepper into small pieces, grate the onion and mix them together. Take the stones out of the avocados, scoop out the flesh with a spoon and mash it with a fork or in the food processor, adding the lemon juice to prevent it going brown. If you are using a food processor, leave the avocado in it and add the garlic (pressed first), a few drops of tabasco and the other seasonings, including the chopped chilli pepper (optional). Blend thoroughly. If you are not using a processor, beat the garlic and spices into a little olive oil and pour it over the salad last of all. Mix the avocado mixture with the other vegetables and adjust the seasoning to your taste. If the chillies make things a little fiery, serve on a bed of shredded lettuce or alfalfa sprouts to cool it down!

We once tried to explain to a three-year-old friend that Guacamole might be a little hot for his taste, but undaunted he replied 'Oh, that's OK, I'll blow on it.'

GREEK SALATA

The more authentic you can make this – with Greek olives, Greek olive oil, big Mediterranean tomatoes, fresh oregano – the more delicious it is. Good olive oil is perhaps the stiffest challenge. Good cooks are as fussy about oil as connoisseurs are about wine because there are such great differences in quality. Look for cold-pressed virgin olive oil. This means that it is a first grade oil, extracted from the peeled and stoned olives by gentle pressure. Repeated pressings give second and third grade oils. Choose an oil that is the product of one country only as each country has its own distinct flavour – Greek oils, for example, are generally more 'fruity' than French. Olive oil should be kept in a cool place to prevent it going rancid, but not in the fridge which is too cold and makes it thick and difficult to pour.

4 large tomatoes, ½ cucumber, 1 onion, 2 handfuls of black olives, sheep's or goat's feta cheese or white Stilton, dried or fresh oregano, 2t crushed coriander seeds, 1 clove garlic, 4T olive oil, freshly milled black pepper, kelp or salt substitute

Use a wooden salad bowl for best effect. Rub the inside with the slightly crushed garlic clove. Slice the tomatoes and cut each slice in half. Slice the cucumber and slice or finely chop the onion. Put these in the salad bowl. Add the olives, sprinkle with the oregano, crushed coriander seeds and other seasonings and toss well. Crumble the cheese into the salad or add it in thin strips. Finally sprinkle the olive oil into the bowl and toss lightly.

Green Salads

SPUN SPINACH SALAD

There are several spinach substitutes which taste as good as, if not better than, best quality spinach. One of our favourites is 'perpetual beet' which closely resembles spinach and grows all year round, but tastes less sharp.

Spinach is traditionally hated by force-fed children. Believed by conscientious parents to be the key to 'growing up strong and healthy', its benefits have been highly misunderstood. Spinach *is* rich in vitamin A and startlingly high in iron and calcium, but when served cooked, as it usually is, the oxalic acid in it makes these goodies unavailable for absorption. With raw spinach this problem doesn't occur. If Popeye had eaten fresh raw spinach he might have been Superman!

large bunch of spinach leaves with stems removed, 2 handfuls of button mushrooms (finely sliced), a few red radishes (sliced), 3 spring onions, a handful of cashews or Brazil nuts (coarsely chopped) or a few sunflower seeds

The trick here is to make the spinach look spun. Using a very sharp knife and holding the leaves tightly bunched together, pretend you are cutting paper thin slices of bread. The result will be exquisitely fine green shreds which look and taste delicious. Chop the spring onions finely so that their taste blends with the spinach, and add the other ingredients. The mushrooms should be sliced down their stalks. Toss with a light French dressing with garlic.

FLORENCE FENNEL SALAD

In Italy fennel is often eaten raw at the end of a meal as a digestive. It has a refreshing aniseed taste.

2 fennel bulbs (cut into fine slivers), 1 romaine lettuce (torn into small pieces), ½C coarsely chopped pecan nuts, 1 C ricotta cheese or tofu, 1 T chopped chives, feathery fennel tops to garnish

Discard the outer leaves of the fennel if they are tough and either slice the whole bulbs crosswise or separate them into stalks and then slice them. Mix with the torn lettuce leaves. Dice the cheese or tofu and add it to the salad along with the nuts. Dress with an Italian dressing and serve with a sprinkling of chives and fresh fennel tops (or fennel seeds).

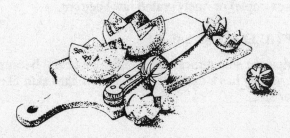

LETTUCE SALADS

Lettuce is a natural tranquillizer. It contains small amounts of a chemical called lactucin, known to induce a state of relaxation when taken in large quantities. Wild lettuce when eaten or smoked is said to have an effect similar to marijuana!

Usually you pile vegetables into a lettuce salad in an attempt to give it flavour. But really one should not bother to use a lettuce at all if it is limp and tasteless. The French prepare wonderful plain lettuce salads using just one kind of lettuce, or an assortment of several, picked the same day. They simply wash and spin dry the leaves, then toss them into a bowl whole, or gently torn, with a little lemon juice and olive oil. So the key to lettuce salads is to find fresh lettuce! If this is not possible you have to do your best to revive the lettuce you have. Place the whole lettuce, roots down, in a basin of cold water and leave in the fridge for about two hours. Then rinse the leaves individually in cold water and spin dry gently. Put them in a polythene bag in the fridge until you need them.

Some lettuces keep their crispness longer than others. The floppy Dutch type seem to be the worst! Go for the naturally crisper varieties if you can – Webb's Wonder, cos, romaine, radicchio (has purple-tinged leaves), and Chinese leaves. They may be more expensive, but they are much more tightly packed, so you get more for your money.

Try adding a few tablespoons of toasted sesame seeds, or thinly sliced water chestnuts or finely sliced white radish to a plain lettuce salad. Or top with grated hard cheese (parmesan for example) or finely grated raw beetroot.

SNAPPY ALLIGATOR SALAD

Avocados are sometimes called 'alligator pears' because a certain variety has a dark greenish-brown hard skin like an alligator's.

1 lettuce, 1 small leek, 2 tomatoes (finely chopped), 1 avocado (finely chopped), a handful of shredded almonds (alligator teeth!), fresh or dry basil, tabasco sauce or cayenne pepper, radish sprouts (optional), lemon juice

Make a bed of lettuce leaves, torn, shredded or whole. Wash the leek well and cut it into long narrow strips about 1 in/2.5 cm long. You can use the green tops almost to the end if they are crisp and juicy. Chop the tomatoes and avocados and mix with the leeks and lemon juice. Pile everything onto the lettuce bed and sprinkle with the shredded almonds (toasted if preferred) and a few radish sprouts. Add the herbs and seasonings to a French dressing and pour over the top.

TASTE BUD TREAT SALAD

This recipe combines all the flavours which your taste buds are able to recognise without the help of your nose – sweet, sour, bitter and salt! (Watercress keeps best if you stand the stems in a glass of water in the fridge.)

large bunch watercress, 6–8 small radishes, 3–4 sticks celery, chopped walnuts

Wash the watercress and remove the thickest stalks. Make radish 'roses' (see page 210) and pop them in cold water for half an hour for the petals to open. Chop the celery finely and coarsely chop the walnuts. Prepare the dressing by combining the following:

4T olive oil, 1T lemon juice, 2t tamari, 1t honey, ground pepper, minced onion (optional)

Put the watercress in a dish and sprinkle the chopped celery and walnuts on top. Decorate with the radish 'roses' and pour the dressing over the top.

WILD GYPSY SALAD

This salad is different every time we make it. Everything depends on the time of year and the herbs and wild edibles available. We gather the ingredients, remove the leaves from the stalks, and nibble each kind of leaf to see how strong a flavour it has, and whether it is mild or dominant. Then we decide on the quantities to use.

Some ingredients could be: dandelion leaves, sorrel, purslane, chickweed, fat hen, lamb's lettuce/corn salad, jack by the hedge, mustard and cress or watercress, thyme, nasturtium leaves, bergamot, chervil, basil, lovage and marjoram. Mix everything with a little citrus fruit and dress with a citrus dressing, or eat plain with a yoghurt dressing.

Mainly-for-winter Salads

Most people say 'Oh yes, I like a nice salad in summer, but I want something hot in winter.' They don't realise that the energy raw food gives you (which is warming in itself) is just as important in winter as in summer. Do they want to be healthy for only half the year? Of course the disadvantage in wintertime is that the selection of fresh vegetables is limited. Nevertheless with a little ingenuity one can prepare some delicious winter salads.

KING OF HADES' SALAD

This salad is dedicated to Pluto, god of the underworld, because its main ingredients are grown underground.

2C grated carrots, ¹/₂C finely grated beetroot, ¹/₂C grated turnip, parsnip, Jerusalem artichokes, kohl rabi, potato, white radish (choose whichever three of these most appeal to you), raisins or diced onion, nutmeg

Mix all your grated root vegetables together and add the raisins or diced onion. Pour a mayonnaise, yoghurt or fermented seed dressing over the top and dust with nutmeg. Nutmeg is very warming and said to make one merry. Add a tub of plain cottage cheese and you have a very substantial main dish.

PERSEPHONE'S TEMPTATION

Zeus proclaimed that Persephone, abducted by Pluto, the King of Hades, could only return to the bright world above if she refused to eat all food served to her in the Underworld. She held out for a while, but finally succumbed to the temptation of a few succulent ruby-red pomegranate seeds. Little wonder!

10–12 young Brussels sprouts, 4–5 sticks celery, 2 pomegranates, 2 or 3 seedless satsumas, sunflower seeds

Discard the outer leaves of the Brussels sprouts and grate the rest finely. Finely chop the celery. Remove the pomegranate seeds (don't use the 'pith' as it is extremely bitter). Peel and slice the satsumas crossways as you would oranges, then separate the sliced segments. Mix all the ingredients together and dress with a yoghurt and honey sauce. Sprinkle sunflower seeds (toasted if you wish) over the top.

Cole Slaws

Cole slaws are great winter salad favourites in our family, but of course the ingredients can be bought all year round. The principal ingredient of cole slaws – cabbage – has many health-giving properties. High in vitamin C, raw cabbage juice is used to treat stomach ulcers. It is also a good blood tonic for people with an iron deficiency and cleanses the mucous membranes of the stomach and intestines. In the past cabbage leaves were warmed and crushed and either laid on the skin to cure such ailments as eczema or used as bandages for wounds and sores.

When choosing a cabbage make sure its leaves are tightly packed, that the head is heavy and firm, and that it still has its outer leaves on. To keep it fresh wrap it in cling film and keep it in the fridge.

ROYAL SLAW

A slaw fit for a king. This recipe includes caraway seeds, useful for dispelling painful stomach wind. Some people chew a small handful before a meal to prevent flatulence and stimulate digestion. (Dill seeds have similar uses.) Caraway seeds are an ideal addition to cole slaws as cabbage tends to cause flatulence in some people.

1–2C grated red cabbage, 1–2C grated white cabbage, plus a few outer leaves of either, ½C grated carrots, 1 stalk celery (finely sliced), ½ red pepper (grated), 1 small onion (optional), raisins, ½t celery seeds, 1t caraway seeds, 1t dill seeds

This salad can be made very quickly in a food processor. Use a medium sized grater rather than a small one, which makes a mushy slaw that swims in its own juices. Grate the ingredients one by one, omitting the onion if you like. Line a bowl with the cabbage leaves and put all the grated ingredients, mixed together, inside them. Scatter with the raisins and seeds. Use a mayonnaise dressing with cayenne in it for bite.

RAINBOW SLAW

This is a nice salad to serve at parties because it is so eye-catching.

Choose half a dozen different vegetables of contrasting colour and shred or grate them, then arrange them in curving rows on a large platter to form a rainbow pattern. Clean your processor grater between the different vegetables or you will get the colours staining each other. Try some of the following in quantities of 1C or so:

red cabbage, white cabbage, green cabbage, carrots, beetroot, turnips, parsnips, swedes, shredded beet greens or spinach

Put a 'pot of gold' of salad dressing at one end of the rainbow!

JUNGLE SLAW

The peanuts really make this one. Peanuts are known to contain large amounts of the amino acid tryptophan. This is a sleep inducer and is also thought to be an anti-depressant.

2C white cabbage (shredded or finely grated), a handful of tender green beans (cut into diagonal slivers), 2 carrots (grated), ¹/₂ onion (grated), ¹/₂ red/yellow/green pepper (finely chopped), 1C unsalted peanuts

Mix all the ingredients together except the peanuts. Make a dressing with peanut oil (if you can find some) and orange juice (see Citrus French dressing on page 240 as a guide). Add a little finely chopped fresh chilli if you're feeling adventurous. Toss the peanuts in at the last moment so that they don't become soggy. For extra flavour toast them lightly under the grill for a few minutes.

SPINACH SLAW

A delicious combination of flavours. Be sure the spinach is really fresh and crisp. If it looks a little droopy, wash it, put it in a bowl of cold water for about half an hour, then drain it and put it in the fridge in a polythene bag to crisp up. Don't forget to remove the tough main ribs from the leaves.

large bunch young spinach leaves, 1C white cabbage, 3 apples (preferably sweet red ones), a large handful of seedless raisins, 1 orange and 1 lemon, juiced

Soak the raisins in the fresh orange and lemon juice for about an hour. Shred the spinach leaves, grate the cabbage, and dice the apples, and put them all into a bowl. Stir in the soaked raisins and juice. Use an oil dressing with a little honey in it.

SMOOTH TURNIP SLAW

For this you will need about 4C of grated or finely diced turnip. Make a sauce by blending ½C sour cream or yoghurt with 1T cider vinegar and 2T honey. Put the turnips in a shallow dish and drop a little fresh parsley on top.

BLOOD-APPLE SLAW

This was our introduction to eating raw beetroot – until we tried it we weren't at all convinced! Actually beetroot is an excellent liver cleanser. Now it is one of our favourite salads.

6 sweet eating apples (grated), 2 small beetroots, orange juice, fresh mint, cinnamon or *nutmeg*

Grate the apples and sprinkle with the orange juice to stop them going brown. Finely grate the beetroots (no need to peel them, just scrub thoroughly), drain off any excess juice, and add to the apples. This salad does not really need a dressing. Serve it with a dusting of cinnamon or nutmeg and a garnish of mint sprigs. As a variation grate a carrot into the mixture too or add some finely chopped celery.

Sprout Salads

You do not have to be at the mercy of your greengrocer or health food shop to make these salads since you can grow most of the ingredients yourself all year round (see Chapter 18). Once you become acquainted with sprouts you will probably want to have some on the go all the time so that you can use them in many of your dishes for their fantastic nutritional value and flavour!

Although sprouts can be added to almost any salad, we often make salads from sprouts alone.

CONFETTI SPROUT SALAD

The more variety of sprouts in this salad the better!

2C alfalfa sprouts, 1C mung sprouts, 1C lentil sprouts (red, green and black lentil for the most stunning colour effect!), 1/2C chick pea sprouts, 1/2C fenugreek or *aduki sprouts, a few radish sprouts, sunflower seeds*

If you have any buckwheat or sunflower greens use a cupful to make a bed for the other sprouts, otherwise shred some lettuce or other greens. Mix the sprouts together and dress with a spicy seed ferment or mayonnaise dressing. Some sunflower seeds soaked overnight are a nice addition. Spoon the sprout mixture onto the bed of greens and sprinkle with a little dulse or some grated cheese and paprika.

SWEET WHEAT SALAD

Sprouted wheat and carrots combined have a surprisingly sweet taste.

2–3C wheat sprouts, 2 grated carrots, sesame seeds or *tofu to decorate*

Mix the grated carrots and sprouts together and garnish with sesame seeds or tofu. You can use a lemon dressing with a little parsley, or stress the sweetness of the salad with a honey dressing and a few raisins.

CRUNCHY PROTEIN SALAD

Together wheat sprouts and chick pea sprouts (like many sprout combinations) provide a complete protein, all eight essential amino acids.

2C wheat sprouts, 1C chick pea sprouts, 1 red pepper, 2 spring onions, fresh parsley

Mix the sprouts together and decorate with thin rings of red pepper. Add the spring onions, finely chopped, and the parsley to a French dressing and pour over the top.

The Arabs make a very similar salad, by omitting the chick pea sprouts and using wheat sprouts only, seasoned with chopped spring onions, parsley and mint, and dressed with lemon juice, olive oil and garlic. Tomatoes and black olives give good contrasting decoration.

ORIENTAL SPROUT SALAD

This is good prepared an hour or two in advance and left to marinate in its dressing.

2C mung bean sprouts (grown to a length of 2–3 in/5–7 cm), 2C mushrooms (thinly sliced), 1C white cabbage (finely shredded), bunch watercress or chives

Mix the mushrooms, sprouts and cabbage together. Place in a shallow dish and pour on the following dressing:

2T olive oil, 2T tamari sauce, 2T lemon juice and a little grated rind, a dash of vinegar, 2t honey, ¹/2t finely grated fresh ginger or ¹/2–1t powdered ginger

Garnish with watercress leaves or chopped chives.

Savoury Salads with Fruit

SUNSHINE SALAD

The fresh pineapple gives this salad a tropical taste. Make sure your pineapple is ripe by pulling out one of its centre leaves; if it comes out easily it is ready to eat.

1 fresh pineapple, 2 carrots, 2 sticks celery, ¹/2 green pepper, a few crisp lettuce leaves, 2 handfuls of sultanas (or raisins), ¹/2t celery seeds, 1t dry mustard mixed with mayonnaise or French dressing

Wash and crisp the lettuce leaves in the fridge. Peel the pineapple (it is not necessary to core it) and cut it into fairly small cubes. Coarsely grate the carrots, and finely chop the celery and green pepper, and add them to the pineapple cubes. Add the sultanas, soaked in water for a few hours to plump them up. Sprinkle with celery seeds and serve on a bed of crisp lettuce leaves. Serve with a piquant mustardy mayonnaise or French dressing.

SPICED WALDORF SALAD

4–6 apples (red or green or both), 3 stalks celery, 2 handfuls of raisins, 2 handfuls of walnuts or pecans, 1 handful of toasted pumpkin seeds or cubes of mature Cheddar cheese, lemon juice, ¹/2C mayonnaise, spices (mixed cinnamon, nutmeg and allspice)

Quarter the apples and cut out the cores, then dice and toss them in lemon juice. Slice the celery stalks diagonally. Break the walnuts or pecans in half. Mix the apples, celery and nuts together, adding the pumpkin seeds or cheese and the raisins. Mix the spices, about 1½t altogether, into the mayonnaise and pour over the salad. Toss well. Eat as it is or served on a bed of shredded Webb's Wonder.

As a variation try making Curried Waldorf Salad, substituting ½t curry powder for the other spices.

ORANGE ORANGE SALAD

A surprising combination . . . but it works!

4 carrots, 1–2 C white cabbage, 6 oranges, 2 handfuls of raisins or small seedless grapes, 4t sesame seeds

Coarsely chop the carrots. Juice four of the oranges and blend the juice with the carrots until you have a smooth mixture. Finely shred or grate the cabbage and put it in a bowl with the raisins or grapes. Pour the carrot mixture over it and lightly mix with a fork. Sprinkle with the sesame seeds and garnish with the two remaining oranges, peeled and sliced.

Salad Sprinkles

Whatever your salad, whether an elaborate mixed salad or a simple lettuce salad, it will almost certainly be improved by Salad Sprinkles. You dredge these tasty little extras on the top or serve them in a separate dish so that people can help themselves. Some sprinkles are cooked, but we have included them because they are delicious and still nutritious.

Sunflower, sesame and pumpkin seeds – any or all of them – whole, ground or toasted

Fennel, celery, poppy, caraway, dill, cumin seeds – plain or toasted

Mustard and cress

Wheatgerm

Chopped wheat grass – full of good vitamins, enzymes and helpful bacteria

Ground nuts

Fresh herbs

Seaweed – either soaked and cut into slivers, or dry and crumbled

Flower petals – marigolds for example, or waterlilies

Soya nuts – these are wonderful! You simply bake soya sprouts (sprinkled with garlic powder or vegetable bouillon) in a moderate oven for about 15 minutes, or until brown and crunchy.

Wheat and barley roasts – these have a lovely sweet flavour. Bake wheat and/or barley sprouts on a baking sheet as for soya nuts.

Hard-boiled eggs – since egg whites should not be eaten raw, it is probably best to eat your eggs cooked. They are delicious boiled and then finely grated (in the food processor if you wish) and mixed into salads, particularly spinach or cabbage.

Finely grated beetroot – adds colour to anaemic-looking salads!

Grated hard cheese – parmesan for example

Tofu – cut into slivers or small cubes

SALAD DRESSINGS AND DIPS

The dressing you choose for a salad is as important as the salad itself. A good dressing can make a tasty dish of even the plainest salad. Most people tend to stick to French dressing or salad cream-type mayonnaise because these are the most familiar, but do you know about fermented seed dressings? or carrot dressing? or tahini mayonnaise? In fact, you can prepare strikingly different salads using exactly the same salad ingredients but changing the dressing.

Egg Mayonnaises

Mayonnaises are best for cole slaws, sprout salads and finely chopped mixed vegetable salads. They are thick and creamy and give smoothness and body. Diluted with a little water they are a complement to leafy salads too.

There seems to be a certain mystique attached to mayonnaise making. Some people say that it is difficult to get mayonnaise to 'take' unless you know the secret. There are no secrets. It is perfectly simple, particularly if you use an electric blender. We find that olive oil makes a rather strong-tasting mayonnaise, so we sometimes use lighter cold-pressed oils instead, sunflower or safflower for example (see our comments about oils on page 177). We have tried several ways of preparing a basic mayonnaise, and this is the one we find works best:

BASIC EGG MAYONNAISE

2 egg yolks, 2 T cider vinegar or *lemon juice, 1 t mustard powder* or *fresh French mustard, 1 t honey, pepper, and vegetable bouillon powder, 1/2 pint/3 dl oil (olive, sunflower, safflower)*

Put all the ingredients except the oil into the blender and process at top speed for about 45 seconds, then slowly trickle the oil through the hole in the top of the blender in a thin continuous stream. When the mayonnaise starts to take you will know without looking because the sound of the blender will change. This method really does work, probably because pre-blending all the ingredients raises them to room temperature, important if your eggs come straight from the fridge.

This recipe makes just over a cupful of mayonnaise. If put in an airtight jar in the fridge it keeps for a few days (bought mayonnaise only keeps so well because it is loaded with preservatives).

VARIATIONS ON BASIC EGG MAYONNAISE

Once you have made a batch of basic mayonnaise you can branch out with lots of interesting flavours and spices.

Garlic Mayonnaise   Add a small crushed clove of garlic
Curry Mayonnaise   Add ½t curry powder and a dash of nutmeg
Horseradish Mayonnaise   Add 1t finely grated horseradish
Hot Mayonnaise   Add a dash of tabasco sauce *or* ¼t chilli powder
Paprika   Add 1t paprika for a beautiful pink colour
Mint   Blend in a handful of clean, de-stalked fresh mint leaves
Cheese and Onion   Add a little grated parmesan or hard Cheddar cheese and some very finely grated onion, chopped spring onions or chopped chives
Herb Mayonnaise   Add a small handful of your favourite fresh herbs *or* a tablespoon of mixed dried herbs (parsley, oregano, thyme, tarragon, basil)

Tahini Mayonnaises

Like egg mayonnaise, once you know the basic ingredients and method there are lots of delicious variations you can try.

BASIC TAHINI MAYONNAISE

½C tahini, juice of 1 large lemon, approximately ½C water

Combine the tahini and lemon juice in the blender at

medium speed. Add the water a little at a time to get the consistency you want. This makes just over a cupful of mayonnaise. We use the thick end of a chopstick for scraping the mayonnaise out from under the blade of the blender. Keep the mayonnaise in an airtight jar in the fridge. Ideally it should be used the day you make it, but it will keep up to five days.

VARIATIONS ON TAHINI MAYONNAISE

Plain tahini mayonnaise is a little bland on its own, but some delicious variations can be made.

Small Seed Tahini Mayonnaise Add caraway *or* dill seeds, 1T cider vinegar, some finely grated lemon rind, a little honey and some finely grated onion if desired. Sprinkle a few whole sesame or poppy seeds into the dressing just before you serve it.

Mexican Pepper Mayonnaise Add 2T finely chopped red and/or green pepper, 1T finely minced onion, a pinch of cayenne pepper, a little mustard, and ¼t vegetable bouillon powder.

Herbal Mayonnaise Add half or a whole clove of garlic (crushed), fresh herbs if possible (chives, basil, chervil, parsley, lovage, rosemary) finely chopped, 1t vinegar and 1t honey.

French Dressings

Oil dressings are especially good for leafy salads such as lettuce and spinach. With the right seasonings – mustard and herbs for example – they can be very flavourful and not at all the 'plain oil and vinegar dressing' most people know.

BASIC FRENCH DRESSING

¾C oil, ¼C lemon juice or cider vinegar, 1t whole-grain mustard (Meaux is our favourite) or mustard powder, 2t honey, vegetable bouillon powder and pepper to season, a small clove of crushed garlic (optional)

Combine all the ingredients in the blender or put them in a screw top jar and shake well to mix. Some people like to thin the dressing and make it a little lighter by adding a couple of tablespoons of water.

VARIATIONS ON FRENCH DRESSING

Here are some suggestions for dressings using the French dressing above as a base.

Rich French Dressing   Add 1T tamari, a finely chopped scallion and a dash of cayenne pepper

Wine Dressing   Add 1T red or white wine. White is especially good for salads containing fruit, and red for cabbage salads

Herb   Our favourite combination of herbs for this is: marjoram, basil, thyme, and dill or lovage, about 3–4T in all if they are fresh and finely chopped, or 2t if they are dried

Citrus   Use ¾C sesame oil, the juice of half a lemon and one orange and 1T cider vinegar instead of the first three ingredients given for the basic French dressing. Add 1t grated orange peel and ½t grated lemon peel (scrub the fruits first!), a pinch of nutmeg and 1t chervil. Put all the ingredients in the blender and combine until smooth.

Spicy Italian   To the basic French dressing above add a dash of red wine or tamari, 2 ripe peeled tomatoes, 1T finely chopped onion, garlic, ½t oregano and basil, and some powdered bay leaf. Blend all the ingredients well.

Seed and Nut Dressings

These are particularly 'warming' served over winter salads, and they make a sprout or mixed vegetable salad more substantial. There are two types, fermented and unfermented. Fermented dressings take about 6–12 hours to culture, depending on the temperature. They have a taste all of their own, sweet and tangy. As with most dressings once you know how to make the base you can add herbs and spices to suit your taste.

Nuts and seeds can be fermented separately to make

sauces, but we prefer to ferment them together. Two of our favourites are sunflower with almond and sunflower with cashew nuts.

SUNFLOWER AND CASHEW DRESSING

This dressing is good served over grated carrots and other root vegetables.

½C cashews and ½C sunflower seeds, 1C water, 1t yeast extract or vegetable bouillon powder

Grind the nuts and seeds as finely as possible. Add the water, then the yeast extract *or* vegetable bouillon and mix well. Put the mixture into a bowl, cover with a tea towel, and place in a warm spot for about eight hours (or overnight). After a couple of hours give the sauce a good stir. The ferment should taste sweet and pleasant. If it is too thick, simply thin it with water; if it tastes 'off' you have overfermented it. Season lightly with a few fresh herbs before serving.

SUNFLOWER AND SESAME SEED DRESSING

This is an 'instant' fermented dressing made with rejuvelac which is itself a ferment (to make rejuvelac see page 200).

½C sunflower seeds (preferably soaked overnight in water), ¼C sesame seeds, juice of one lemon, 1t tamari, 1C rejuvelac, basil or sage and vegetable bouillon to taste

Grind the sesame seeds finely, then add the sunflower seeds and re-process (the sesame seeds tend not to get ground up unless they are processed by themselves). Add the rejuvelac, herbs and bouillon and blend well. You can use this one right away.

ITALIAN PESTO

A delicious sauce this, particularly good served over alfalfa sprouts or a simple lettuce salad. Pine kernels or pistachios give it an authentically Italian taste. If you cannot get either, or if they are too expensive, substitute almonds or pecans.

1C pine kernels (pignolia nuts) or pistachios, a handful of fresh basil leaves, a little parsley or oregano, 1/2C olive oil, 1/2 clove crushed garlic, a little grated parmesan or sardo cheese if desired

Blend or process the nuts and gradually add the oil. Blend in the herbs (remove the stalks and use only the leaves), garlic and cheese, and serve.

Vegetable-based Dressings

These make a refreshing change from oil dressings and mayonnaises, but they should be prepared just before serving because vegetables quickly lose their flavour and nutritional value once they have been finely ground. Very tasty and colourful dolloped over Dish Salads.

DRACULA'S DELIGHT

1 small beetroot, 1C toasted or raw sunflower seeds, approx. 1C water, juice of 2 lemons, a little grated lemon rind, 2T tamari, cayenne, basil, garlic and vegetable bouillon powder

Scrub and grate the beetroot, put it into the blender with the rest of the ingredients, except the herbs and bouillon, and process finely. Now add seasoning to taste.

TOMATO TREAT

You can use tinned tomatoes for this but they are not as flavourful or as health-promoting as fresh ones.

4–5 large tomatoes, 1/2C almonds, 1/2 avocado, juice of 1 lemon, 1 spring onion, small clove garlic, a handful of fresh basil leaves, a dash of tamari or vegetable bouillon powder

Peel the tomatoes, then blend well with the rest of the ingredients. Dilute to the required consistency with a little water. Very good served over Dish Salads, or over green or tomato salads.

CREAMED CARROT

1–2 carrots, 1 C carrot juice or water, 2 T olive oil, ¼ C walnuts, a handful of fresh parsley, 1 t dill, small clove of garlic (optional), vegetable bouillon to season

Chop the carrots and blend them with the walnuts, oil and carrot juice or water. Add the parsley, dill, garlic and bouillon and blend again. This dressing is delicious with a little cream cheese or fresh cream added to it . . . it makes it that bit more creeeamy!

AVOCADO DRESSING

Simply substitute the flesh of a ripe avocado for the oil in any of the French dressing recipes and add a little water or vegetable juice to give the consistency you want.

OLIVE DRESSING

Blend 1 C olive oil with the juice of one lemon, half a dozen chopped pitted black or green olives, some fresh basil and a little cayenne pepper. This will keep in the fridge for at least a week.

Yoghurt Dressings

Yoghurt makes a nice light dressing and is particularly good for weight watchers. Here are two of our favourites:

GREEN GODDESS DRESSING

This is very refreshing and ideal for summer salads.

1 C yoghurt, 1 handful of fresh herbs (mint, basil, lovage, parsley), 1 t French mustard, juice of ½ lemon

Blend all the ingredients until the herbs are finely chopped.

SPICED YOGHURT DRESSING

A stronger dressing with more 'zip' for a plain or bland salad. Simply blend all the ingredients together well.

1C yoghurt, 2T oil, 1T cider vinegar, 1T tamari, 1T finely minced onion or scallion, 1/2 clove garlic, 1t molasses, 2t honey, a pinch of cayenne pepper, a little dill, parsley, thyme, sage and celery seed

Dips

Dips make a scrumptious accompaniment to a plate of crudités or spooned into the centre of Dish Salads. Some of the thicker ones are thick enough to serve as pâtés with narrow strips of Essene bread.

HOUMOUS

A raw version of the famous Greek appetiser which we find even more delicious than the cooked original.

1C chick pea sprouts (about 1 in/2.5 cm long), juice of 1 lemon, 2T orange juice, 1 clove garlic, 2T tahini, chives, vegetable bouillon

Blend the chick pea sprouts very finely in the food processor. Add the lemon juice, orange juice, tahini, garlic and vegetable bouillon. Mix well and serve sprinkled with chopped chives. To make thin houmous dressing for salads – a delicious change – simply thin with yoghurt or water.

COOL CUCUMBER DIP

1 small cucumber, 3/4C yoghurt, squeeze of lemon or dash of vinegar, 1T minced onion, 1t honey, 1 clove garlic (optional), fresh mint, pepper, salt substitute

Peel and grate the cucumber and drain off any extra juice (you can use it in a drink). Mix with the yoghurt, lemon juice or vinegar, onion, honey and garlic. Finely chop a few mint leaves and add. Season and serve in a dish with sprigs of fresh mint.

CARROT DIP

This is made with cow's milk products, but you could quite successfully substitute soft goat's cheese or even seed cheese and yoghurt.

½C sour cream, ½C cream cheese, 1 large carrot, paprika, 2 spring onions

Blend the cream cheese and sour cream with the grated carrot for a few moments. Scoop into a bowl and sprinkle with finely chopped spring onions and paprika.

CURRIED AVOCADO DIP

A hot favourite in our house!

1 or 2 avocados, 1C orange juice, 1t curry powder, 2t vegetable bouillon, a few lovage leaves, parsley, fresh basil or marjoram, 1 clove garlic

Blend the avocado flesh with the orange juice in the food processor and add the seasonings to taste. Adjust the amount of orange juice to get a thick dip or a thin dressing.

TARAMASALATA

There are many variations of this famous Greek dip, some including potato and bread. This recipe is simple, but nonetheless very tasty.

½C smoked cod's roe, ¼C olive oil, 2T water, 1 clove garlic, dash of tabasco sauce, freshly ground pepper or paprika

Skin the roe and process it with the pressed garlic until smooth. Gradually add the oil and water while the food processor is running. Season with a little tabasco and freshly ground pepper or paprika.

THOUSAND ISLAND DIP

This contains hard boiled eggs, but is too delicious to leave out!

1C egg mayonnaise (see page 237), 1t Meaux mustard, ¼ red pepper, 2 hard boiled eggs, slice of beetroot, 4 green olives or 1 gherkin, fresh parsley

Cut the red pepper into tiny pieces, discarding the seeds. Finely chop or grate the eggs. Stone the olives and chop finely. Cut the beetroot slice into tiny squares. Combine the mayonnaise and mustard and add the 'islands' (red pepper, egg, olives and beetroot). Serve with fresh parsley sprigs.

CROQUETTES, PATTIES AND LOAVES

Very often when people start eating a raw diet they complain that salad and fruit simply don't satisfy them, that they need meat or bread to really fill them up. The dishes in this section are designed to do just that, to supplement the main course of your meal (a large salad) with extra protein and calories in the form of nuts, seeds and sprouts to give you that 'full' feeling. After eating lots of raw foods for a while you will find that your 'hunger pangs' decrease, that you no longer want to eat a lot of heavy foods such as nuts, and that a large salad for a main meal will be quite enough to leave you feeling satisfied, but not bloated.

Loaves

SANDSTONE LOAF

This dish has a beautiful pink/orange colour and is very easy to prepare with a food processor.

6–8 carrots, 3–4 sticks celery, ½C almonds or peanuts, 2T tahini, ½ onion, juice of ½ lemon, a handful of fresh parsley or 1T dried, 2t vegetable bouillon powder

Wash the carrots and celery. If the celery is stringy peel away the tougher fibres. Roughly chop the carrots and celery and put into the food processor. Homogenise thoroughly, adding the lemon juice, and put into a separate bowl. Now grind the nuts as finely as possible. Add them to the carrot and celery mixture, and stir in the tahini, finely chopped onion and parsley. Pack into a bread tin. Garnish with parsley leaves and serve.

PHONEY PHEASANT LOAF

The cranberry sauce finishes this recipe off nicely. Cranberries have an affinity for 'game'. The American Indians used to mix their venison with cranberries to preserve it for the winter. Nowadays people drink cranberry juice to help fight off infections. You could substitute blackcurrants or redcurrants if you cannot get cranberries.

4–6 sticks celery, 1 scallion or 2 spring onions, 1C cashew nuts, 1C pumpkin seeds, ½C brazil nuts, fresh parsley, 1t sage, 1C cranberries (blackcurrants/redcurrants), honey

Grind the nuts and seeds in the food processor. Add the chopped celery and scallion and homogenise (for a crunchier loaf chop them finely and combine without processing). Add the herbs and mix well. Turn into a loaf tin. Make a sauce with the berries by blending them, straining most of the juice off (use it in a drink!) and adding honey to taste. Spread the berries over the top of the loaf and garnish with parsley.

FERMENTED SEED LOAF

This loaf needs to be fermented for 24 hours so make it at least a day before you want it.

½C almonds, ½C sesame seeds, 1C chopped cauliflower or broccoli florets, 4 mushrooms, 2 sticks celery, 2T tamari, 1 clove garlic, basil, parsley, 1t caraway seeds, ½–1C water or rejuvelac, radish slices to garnish

Finely grind the nuts and seeds. Add the seasonings – tamari, chopped garlic, basil and parsley, and the caraway seeds – and the water or rejuvelac. Finely chop or grate the cauliflower or broccoli and dice the celery and mushrooms. Mix all the ingredients together and pack into a bread tin. Cover with a tea towel and leave to ferment for 24 hours in a warm place. Add radish slices just before serving.

Croquettes and Patties

SPRUNG SPROUT CROQUETTES

A good way of using up sprouts if you find you have grown too many.

A handful each of mung, lentil, fenugreek and alfalfa sprouts, a few radish sprouts, 1 scallion or 2 spring onions, ½C sunflower seeds, ½C cashew nuts, 1–2T tamari, 1T dried oregano, 1t vegetable bouillon powder

Grind the cashews and sunflower seeds finely, adding the finely chopped scallion, herbs, bouillon powder and tamari. Add the sprouts and process for just a few seconds so that they retain their crunchiness. Form the mixture into croquettes or balls, chill and serve.

CHICK PEA CROQUETTES

1C chick pea sprouts, ½C sunflower seeds or peanuts, 3–4 carrots, 1 shallot, 1 egg yolk, 1T tahini, juice of ½ lemon, ¼t cayenne pepper, cumin, poppy or sesame seeds, fresh parsley

Homogenise the chick pea sprouts, sunflower seeds or peanuts, egg yolk, tahini, lemon juice and cayenne pepper in the food processor. Finely grate the carrots and shallot and add to the chick pea mixture. Season with a little cumin. Form into croquettes and sprinkle with poppy or sesame seeds. Serve on a bed of lettuce or alfalfa sprouts and garnish with fresh parsley.

WHEAT PATTIES

2C wheat sprouts, 1C mushrooms, ¹/2 green pepper, 2 spring onions, 2 tomatoes, 1–2T French dressing, vegetable bouillon powder to season

Combine the wheat sprouts and mushrooms in the food processor, blending them lightly so that they don't form a paste. Finely chop the green pepper, spring onions and tomatoes and stir into the wheat and mushroom mixture. Add as much dressing as you need to give a mouldable consistency and season with bouillon powder. Form into balls and flatten on a plate. You can serve the patties with a barbecue-type sauce made in the blender by combining:

¹/2C water, 1T vinegar, 1T Worcestershire sauce, juice of 1 lemon and half the grated rind, 2T honey, 2 peeled tomatoes, a dash of tabasco sauce

CARROT YOGHURT PATTIES

These are an old favourite of our family and very simple to make.

6–8 carrots, 3 sticks celery, 3 spring onions, 1C sunflower seeds or ground mixed nuts, ¹/2C yoghurt, wheatgerm, lemon juice, vegetable bouillon powder, basil and pepper

Grind the sunflower seeds or nuts finely or coarsely depending on how crunchy you want the patties to be. Grate or homogenise the carrots and finely chop the celery and spring onions. Mix everything together in a bowl. Make a sauce by mixing the yoghurt, lemon juice, bouillon powder, basil and pepper together. Make a well in the centre of the vegetables and pour it in. Add the wheatgerm to get a mouldable consistency (you will need more wheatgerm if you homogenised rather than grated the carrots). Form the mixture into balls and flatten. Eat plain or dusted with chopped parsley or toasted sesame seeds.

DESSERTS AND DESSERT TOPPINGS

Many people give up eating dessert in order to 'watch their weight' or simply because they feel full to bursting point at the end of a rich meal. We think this is a shame, for dessert can be the crowning enjoyment to a meal. Desserts need not be stodgy, rich and fattening. The recipes in this section are mostly made with fruit without added sweetening. They leave you feeling satisfied and energised, not weighed down. Also, some fruits contain enzymes which aid digestion and are therefore an ideal and refreshing way to end a meal. The recipes we have included range from the elaborate to the very simple.

Fruit Salads

SATIN SALAD

This is made mainly from soft fruits, which give it a smooth texture. The most common type of peaches are yellow, but look out for the white variety. They are a pale creamy colour tinged with pink and have a beautiful aroma like fruit blossom. When choosing peaches make sure they are not bruised or green. Unlike other fruits, peaches do not ripen after they have been picked, they merely soften and begin to lose their flavour.

2 bananas, 4 peaches or *nectarines, 1 pear, 2 sweet plums* or *a few cherries, 1 C seedless green grapes* or *1 C soaked raisins, desiccated coconut (optional)*

Slice the plums or halve the cherries, dice the peaches, and peel and dice the pear. Put them in a bowl and add the grapes, whole or halved. Slice and add the bananas. As a topping blend a cupful of the fruit with a little fruit juice or the soak water from some dried fruit. Pour this over the fruit salad. Alternatively, serve plain or sprinkled with a little grated coconut.

DRIED FRUIT SALAD

This is particularly nice in winter when your choice of fresh fruit is limited. The recipe contains figs which are associated with altered states of consciousness, probably due to the high amount of the amino acid tryptophan in them!

A handful each of prunes, dried apricots, dried figs and raisins, 1/2C pine kernels (or blanched almonds or pecans), 1T orange flower water, 1T rose water, 2C water

Place the dried fruit in a bowl of water, cover with a tea towel and leave overnight in a warm place. Remove the prune stones and chill in the refrigerator. Before serving stir in the rose and orange water and the nuts (pine kernels are best, but sliced almonds or pecans are good standbys).

FRESH CITRUS SALAD

When buying grapefruit, avoid those which feel light or have puffy skins – they are likely to be dry and pithy inside. Look out for pink/ruby red grapefruit – they are sweeter than the yellow ones and twice as delicious.

2 oranges, 1 grapefruit or 3 tangerines, 1 small pineapple, fresh mint to garnish

Peel the oranges and the grapefruit or tangerines. Slice them according to the segments, removing each segment from its envelope of skin. Now slice each segment across to give small bite-size pieces. Scoop the fruits and their juices into a bowl. Peel the pineapple, dice it and add to the citrus fruit. Serve garnished with a little fresh mint.

ORCHARD SALAD

A beautiful looking salad which is a combination of subtle reds and pinks.

2 large red apples, 1 pear, 1C cherries, 1 punnet strawberries, 1 banana, ¹/2 lemon, mint sprigs to garnish

Dice the pear, apples and banana and sprinkle them with lemon juice. Slice the cherries in half and remove the stones. Slice the strawberries lengthwise. Mix all the fruits together and decorate with mint sprigs.

Stuffed Fruit Salads

MULLED STUFFED APPLES

The best apples to buy are those which have escaped being treated with pesticide, but they are very hard to find. Look for signs that the skins have been damaged by insects – that tells you that they have been grown without pesticides. Despite appearances they are better than ones with un-blemished plastic-looking skins. Most of the nutritional value of an apple lies in its skin, or just below it, so wash apples well but don't peel them. Softish apples are best for this recipe as their insides have to be scooped out.

4 large apples, 1C grape juice or red wine, juice of ¹/2 lemon, ¹/2–1C blanched almonds, ¹/2C dates or raisins, 1t cinnamon, 3 cloves, ¹/2t nutmeg, 2 crushed white cardamom pods, ¹/2t allspice

'Mull' the grape juice or wine by putting it with the spices in a bowl and leaving for at least an hour. Discard the cloves and cardamom and blend with the almonds. Slice the tops off the apples and keep them. Remove the cores, saving small pieces to plug the bottoms. Scoop out the apple pulp leaving a shell about ¹/4–¹/2 in/ 1 cm thick. Lightly blend the pulp with the juice and almond mixture until smooth, adding a squeeze of lemon juice. If the mixture is not thick enough, add a few more ground almonds. Chop the dates or raisins and mix with the apple stuffing. Fill the apple shells and replace the 'lids'.

Using the same method you can make apples stuffed with apple sauce and blackberries. Blend the apple pulp with a little lemon juice, honey and spices, then combine it with the blackberries and spoon into the apple shells.

STUFFED PINEAPPLE

1 pineapple, 1 orange, 1 mango or pawpaw, 1C raspberries or strawberries, 2 figs (fresh or dried ones soaked), coconut milk (optional), dried coconut to garnish

Slice the pineapple in half lengthwise, and remove the flesh from each half, leaving a ½in/1 cm shell. Dice the flesh and mix it with the sliced orange, mango (pawpaw), and raspberries (halved strawberries). Finely chop the figs and add. Mix all the ingredients with the coconut milk, if desired. Spoon the mixture into the pineapple shells and sprinkle with dried coconut.

ALL MELON SALAD

People who follow a strict food combining regime insist on eating foods of the same group together. Of melons it is said 'Eat alone or leave alone'. Here is a salad which combines three different types of melon and makes a delicious summer breakfast or appetiser.

1 small watermelon, 1 cantaloupe, 1 honeydew, honey

Cut the watermelon in half lengthwise, or if you are more ambitious cut it in half using the zig-zag technique described on page 210 for making Tomato Lilies. Scoop out the watermelon pulp and dice all but a cupful of it – this you can blend or juice to make a drink. Slice the other melons in half and scoop out balls with a melon scoop, or cut in cubes while still in the skin and scoop out the pieces with a spoon. Mix all the melon balls or cubes together and fill the watermelon 'baskets'. Drizzle with a little flowery honey (acacia or lavender) and decorate with fresh flowers (marigolds, elderflowers, waterlilies) and mint leaves.

Fruit Blends

These are an alternative to fruit salads and are particularly good for babies or elderly people for whom chewing is a problem. You can combine almost any fruits, but soft tropical fruits and berries are especially delicious.

SUNSHINE BLEND

½ pineapple, 1 mango or pawpaw, 1 seedless orange, 1 peach, 2T dried coconut, honey

Peel and roughly chop the pineapple, mango and orange. Halve and stone the peach. Combine all these ingredients in the blender until smooth. Sweeten with a little honey if you like. Spoon into frosty chilled dessert glasses. Serve with a little coconut. As a variation use two bananas instead of the mango.

SPICED DRIED FRUIT BLEND

Good for winter evenings when few fresh fruits are available. The spices help to take the edge off the winter chill.

2C dried fruit (apricots or peaches), 2C water, 1 lemon, 1 orange, several allspice or juniper berries, 2 sticks cinnamon, ½t finely grated nutmeg, honey to sweeten

Finely slice the orange and lemon (unpeeled) and place in a bowl with the spices. Add the dried fruit, cover with water and soak overnight. Remove the orange and lemon slices and spices, pour the fruit and soak water into the blender and process until smooth. Chill and serve with a dusting of nutmeg.

Quick Single Fruit Dishes

These sweets can be extremely simple, but are just as delicious as a salad with lots of different fruits in it. They make wonderful light breakfast dishes too.

GRATED PEAR OR APPLE

4 pears or *4 apples, ¹/₂ lemon, honey, cinnamon, ground nuts (brazils, almonds, hazelnuts or cashews)*

Finely grate the pears or apples and sprinkle with lemon juice. Place in chilled dessert glasses or dishes and drizzle with honey. Then sprinkle with ground nuts and just a touch of cinnamon.

JAPANESE ORANGES

Somehow oranges never tasted better! The secret here is how you cut them.

4 large oranges, mint sprigs, cocktail sticks

Slice the bottoms and tops off the oranges, the bottoms so that the oranges do not roll around and the tops so that they can be used as 'lids' later. Using a sharp 'grapefruit' knife cut all the flesh out of the oranges, keeping as close to the peel as possible, cutting through from the top first and then through from the bottom. The top opening needs to be wider than the bottom because you are going to remove the flesh inside in one piece! Slice the flesh into four segments lengthwise, then slice the segments in half crosswise. Put them back in their orange peel shells, replace the 'lids', and chill for an hour or two in the freezer. Serve on small plates decorated with mint sprigs. Skewer a cocktail stick with a sprig of mint into the top of each orange. You eat the segments with the stick.

EXOTIC GINGERED FRUIT ·

Some exotic fruits are best eaten separately to appreciate their subtle flavours. Try mangos, lychees, pawpaws, persimmons, kiwi fruits and so on sliced and sprinkled with a little lemon juice, powdered ginger and honey.

STALKED STRAWBERRIES

One of our all time favourites. First, we wash large strawberries with long stems still attached, and place them in a bowl. Then we fill another bowl with thick yoghurt or sour cream, and another bowl with raw sugar, and place them on the table. Everyone dips their own strawberries into the yoghurt, then into the sugar. Delicious!

Pies and Tarts

There are many possibilities here. But how on earth, we hear you ask, do you make pie crust without flour or even crumbled biscuits? Simple, you use dried fruits and nuts rendered down in the blender and pressed into dishes and tins to form the case for the filling. Tarts are simply mini pies made in individual dessert dishes using the same technique and perhaps chopping the fruit for the filling a little finer. Pies are made in ordinary pie tins and served by the slice. Serve cream, yoghurt or honey with them, or any of the toppings on page 263.

CRUSTS

Our favourite raw pie base is made from dates and almonds, but any dried fruit and nuts can be used.

1 C almonds, 1 C dates, 3 T honey, water if needed

Grind the almonds and dates as finely as possible in the food processor. Add the honey and a little water, if needed, to make the mixture bind. Press into a pie dish or divide into four and press into four small shallow dessert dishes to make tarts. As a variation you could add two or three tablespoons of desiccated coconut or rolled oats, or a tablespoon of tahini. You can also spice the crust with powdered cinnamon or allspice.

FILLINGS

The simplest fillings are sliced or chopped fruits such as plums, peaches or apples, sprinkled with a little orange or lemon juice, drizzled with honey and dusted with cinnamon or powdered allspice. But experiment with your own ideas. You will need about 1–1½ pieces of fruit for each person.

Mince Pie Filling 1C raisins and 1C dates soaked for several hours, 3 apples, honey, cinnamon and nutmeg to taste. Grate or process the cored apples finely and process the dates. Stir in the raisins and other ingredients and spoon into the crust. (This filling is delicious eaten on its own, topped with yoghurt.)

Blended fruit freezes such as banana and strawberry (see page 259) are also good pie fillers.

Apple and blackberry the apples grated, and mixed with lemon juice and honey, also make a traditional and delicious filling.

Cakes

Raw cakes are every bit as tempting as cooked cakes for birthdays and special occasions. The first two recipes contain grain (oats), but the last one does not, and is therefore recommended if you are cooking for someone who has an allergy to grains.

SPICED FRUIT CAKE

Dried fruit: ¼C raisins, 4 peaches (or 6 apricots), 4 pear halves, 4 figs, 4 dates (pitted); ½C mixed nuts and seeds (walnuts and sunflower seeds for example), 3T dried coconut, 1 banana, 1 orange, 1 lemon, ¼t each of cinnamon, nutmeg and allspice, vanilla essence, 1C rolled oats

Put the seeds, nuts, coconut and dried fruit in the food processor and grind coarsely. Turn into a bowl. Blend the banana with the spices and a few drops of vanilla essence, and add ½t orange rind and ½t lemon rind finely grated. Juice the orange and lemon and keep separate. Mix the banana and the dried fruit/nut mixture together in a bowl and add the oats. Stir well and add some juice to give a firm binding consistency. Spoon the mixture into a cake tin with a removable base, or simply shape the mixture into a loaf and wrap it in wax paper. Refrigerate for a couple of hours. Serve the cake in thin slices – it is rich!

CARROT CAKE

The cooked version of this is a favourite in America. Carrots really do make an excellent ingredient for a cake because they are quite sweet.

½C almonds, ¾C rolled oats, ½C wheatgerm, ½C dried coconut, ½C raisins and ½C dates, 3C finely grated carrots, juice of ½ lemon, 4T honey, 3T sesame oil, 1t vanilla essence, 1t cinnamon, 1t allspice, water

Finely grind the almonds and mix them with the rolled oats, wheatgerm and coconut. Soak the raisins and dates in warm water for about 10 minutes (or even better overnight) and blend them in the food processor with the honey, oil, vanilla, spices and two tablespoons of water. Finely grate the carrots, add the lemon juice, and mix into the oat and almond mixture. Make a well in the centre and pour the blended raisin and date mixture into it. Stir well. Pack into a loaf tin and cover with cling film. Refrigerate for a couple of hours. Serve sprinkled with raisins or dried coconut.

CAROB AND APPLE CAKE

3 apples (red ones are nice), 1C sunflower seeds or a 2:1 mixture of sunflower and sesame seeds, 1C carob powder, ½C dried coconut, ½C dates, ½t vanilla essence, 1t allspice, apple slices or strawberries to garnish

Grind the seeds very finely and mix with the carob powder, coconut and finely chopped dates. Core, then finely grate or homogenise the apples. Add them to the other ingredients and stir well. Form the mixture into a loaf or log shape and refrigerate for a couple of hours. Serve sliced with an apple and/or strawberry slice decorating each piece.

Ices

When food is frozen, as in the preparation of sorbets and freezes, the enzymes in it are temporarily inactivated but they are not destroyed. When you eat it and it warms up inside you, the enzymes become active again and behave just as beneficially as those in all raw foods. These raw sorbets and freezes are more delicate in flavour than the kind you buy. We sometimes sweeten ours with a little acacia honey or light unrefined sugar as fruit tends to lose some of its sweetness when frozen.

FRUIT FREEZES

We make these simply by blending our favourite fruits with a little fruit juice or water (if needed) and freezing. To prevent large crystals forming and to make the mixture lighter and easier to spoon, stir every half an hour or so during the freezing process. Here are a few combinations you might like to try.

Blackberry and Peach
Blend two cupfuls of blackberries with four finely grated peaches and a little honey, then freeze.

Redcurrant and Pear
Blend two cupfuls of redcurrants with three finely grated pears.

Berry and Banana
Blend two cupfuls of blueberries *or* raspberries *or* strawberries with three mashed bananas.

Some fruit freezes give a thickish mousse rather than a thin sorbet. The thicker ones, such as Berry and Banana, are nice simply chilled rather than fully frozen.

SMOOTH ICES

For a really creamy texture, remove the seeds or stones from your fruit, coarsely chop it and put it in the freezer. When frozen blend in the food processor with a little water until smooth, light and creamy.

PASSION FRUIT SORBET

The dessert of the gods! The uglier passion fruit look the better they taste; they should be black, wrinkled and fairly soft to touch.

8–10 passion fruit, 4–6 oranges, 1 lemon, fresh mint leaves, honey to taste

Cut the passion fruits in half and scoop out the flesh. Juice the oranges and the lemon. Mix the juice with the passion fruit flesh and add honey to taste. Pour the mixture into a freeze-proof dish and put in the freezer. Serve decorated with mint leaves. Alternatively, keep the orange shells, fill them with sorbet, freeze, and serve garnished with mint leaves.

ROCKY ROAD BANANAS

These make delicious munchy snacks and are very sustaining. They can be made on sticks (the ice lolly sort) or eaten whole with the fingers. Use ripe bananas, with brown speckles on their skins.

4 bananas, 1/2–1C mixed nuts (almonds, brazils, pecans, hazels, cashews), 1/2C toasted or raw sesame seeds, 1/2C honey, carob powder, dried coconut, dates

Peel the bananas and cut in half crosswise (they are easier to cope with at this length). Put several tablespoons of honey onto a flat plate and roll the bananas in it one at a time. Coarsely grind the nuts and sprinkle them, with the sesame seeds, on another plate and roll the sticky bananas in this mixture. To finish off, roll in carob powder or dried coconut or chopped dates or all three! Skewer onto sticks and freeze.

Mueslis

Most of you are familiar with the cereal-type mueslis sold commercially and eaten with lots of milk for breakfast. While these are quite tasty, they are usually loaded with sugar and, eaten with milk, very hard on the digestive system. Home-made mueslis are much better and can be made to suit your specific taste by adding or omitting various ingredients. They can be served as a dessert with a light meal or eaten as a satisfying energy-packed breakfast.

BASIC MUESLI FOR ONE PERSON

The grain ingredients, one kind of grain or several, should be soaked overnight or else used as whole grains sprouted for about three days.

1–2T oat, wheat or rye flakes or barley kernels or whole grains, 1T mixed nuts (almonds, walnuts, brazils, hazels, cashews), 1T wheatgerm (optional), 1 grated apple, juice of ½ lemon or ½ orange, 1T raisins, ¼C yoghurt, 1t honey or molasses, cinnamon or allspice if desired

Put the cereal flakes or kernels in half a cupful of water to soak overnight. Soak the raisins in the minimum of water in another cup. In the morning put the soaked cereals in a bowl with the yoghurt and add the raisins with their soak water. Grate the apple and mix it with the orange or lemon juice. Stir into the cereals and raisins and top with honey, spices, nuts and wheatgerm if desired.

MUESLI VARIATIONS

Grainless muesli

Use two tablespoons sunflower *or* sunflower and pumpkin seeds instead of the cereals, and soak overnight in the same way.

Creamy Muesli

Use fresh cream instead of yoghurt *or* a little goat's milk blended with a tablespoon of ground cashew nuts. Goat's milk on its own gives a thinner muesli.

Dairyless Muesli

Use fruit juice (apple and grape are particularly good) instead of the yoghurt, and leave out the raisin soak water if the consistency is too thin.

Fruitier Muesli

The addition of a little fresh or dried fruit gives added flavour. You can omit the apple, or use only half a grated apple with a little finely chopped banana, peach, plum, strawberries, raspberries, cherries. Or soak some dried apricots, figs, peaches, dates, prunes or pears with the raisins and a little more water.

Sprinkles

Instead of wheatgerm, or as well, try adding some sesame seeds or dried coconut, or barley/wheat roasts (see page 286).

LIVE PORRIDGE

This is a favourite of the youngest member of our family, three-year-old Aaron. Made from fresh fruit, it is beautifully sweet and light and makes a delicious dessert as well as a breakfast dish. The principle is to combine seeds or nuts together with fresh fruit and a 'sweetener' such as raisins or dates. Pears are especially good, but so are strawberries, apples, mangos, blackberries or pitted cherries. You can mix the fruits or use one kind only. As for the nuts sunflower seeds and cashews are nice, or almonds and sesame seeds, or pecans on their own.

5–6 pears, ³/₄C seeds and/or nuts, ¹/₄C raisins or dates, dried coconut (optional but yummy)

Mince the seeds and nuts together in the food processor until they are well chopped. Then add the rest of the ingredients and continue blending until smooth. Pour into four bowls and top with a little fresh cream, dried coconut or toasted sesame seeds, all optional.

Toppings

Some desserts call out for a topping, but it doesn't have to be cream. Sweetened tahini mayonnaise (see page 238) or yoghurt with honey and spices are two of the easiest. Here are some other ideas.

CHOCOLATE BANANA TOPPING

2 very ripe bananas, 5T carob powder, ¹/₄t vanilla essence, water or apple juice to thin

Blend the bananas and carob powder in the food processor. Add a little water or juice to give the consistency you want. Flavour with a little vanilla essence. Alternatively add a tablespoon of coconut and blend well for flavour.

CASHEW CREAM

1C cashew nuts, ¹/₂C water or orange juice, 1–2t honey, nutmeg

Blend the nuts and liquid in the food processor and add a little honey and nutmeg.

RAW APPLE SAUCE

4 small apples, juice of 1 lemon, honey and cinnamon to taste

Quarter and core the apples, then process with the lemon juice until smooth. Sweeten with a little honey and sprinkle with cinnamon.

BREADS AND WAFERS

The idea of raw bread may not seem appealing to you, but in fact sun-baked or Essene bread is one of the most seductive of raw foods. It has a slightly sweet nutty flavour and is delicious plain or with a dip. Some unbaked bread recipes use plain wheat (or other) flour. We prefer to use sprouted grains because they are easier to digest and more nutritious, also because they taste better. You don't need sunshine to 'bake' the breads either. A radiator top, warm stove surface or even a cool oven will do. The important thing is that you keep the temperature below 120°F, the temperature at which vital enzymes and vitamins become inactivated.

Essene Bread – Basic Recipe

The basic ingredient for Essene bread is whole wheat or rye grains soaked for about 15 hours and then sprouted for two or three days. You are aiming for a dough consistency, the sort of consistency you can roll out without its breaking up. Soaked and sprouted grains already contain plenty of moisture, but if your dough mixture is too dry you can always add a little oil or water. If your mixture is too wet, add wheatgerm to it.

Soak and sprout the grains and grind them as finely as possible in the food processor. Add a touch of oil and form into a ball. Place on a bread board sprinkled with wheatgerm to prevent the dough from sticking. Roll the dough into a thin sheet, as you would pastry, with a rolling pin dusted with wheatgerm. Alternatively press the dough out onto a flat tray or board as thinly as possible without breaking it. Leave the bread to 'bake' in a warm place for 6–12 hours, turning over at half time with a pancake turner.

Variations on Essene Bread

SAVOURY SEASONED BREAD

Add half a cupful of ground mixed vegetables (carrots, onions, celery, peppers, parsley, cress) to the wheat in the food processor and grind well. Season with a little vegetable bouillon powder and a tablespoon of seeds (sesame, poppy or caraway). You can also add a sprinkling of dried herbs.

SWEET BREAD

This can be made with ground dried fruit such as raisins or dates – a quarter of a cupful will do – or a small mashed banana. Add to the ground wheat and blend in thoroughly. (If you use a banana you will need to add about half a cupful of wheatgerm to the dough or it will be too wet.) Spice the dough with a little cinnamon, nutmeg or allspice. Add a little honey too if you want a sweeter bread. Actually wheat sprouts have quite a sweet taste on their own.

MILLET BREAD

For a really nutritious bread, try adding a quarter of a cupful of millet, ground into a fine flour, to the basic recipe. You will need to add a little more oil and water to give the right consistency. If you use millet sprouts instead, leave out the water.

SPROUTED GRAIN CRISPS

These make great snacks and can also be used as 'dip chips' or croutons for soups and salads.

1 C wheat (or other grain) sprouts, 1 T tahini, 1 t vegetable bouillon powder, 2–3 T fresh or 2t dried herbs (chives, parsley, basil, marjoram), sesame or poppy seeds to decorate, wheatgerm

Grind the sprouts in the food processor with the tahini, bouillon and herbs. Roll the dough out with plenty of wheatgerm to a thin sheet and sprinkle with the seeds. Cut into strips diagonally with a sharp knife, then cut across these strips diagonally to make 'diamonds' and then through the diamonds to make triangles. Leave to crisp up in a warm place for two hours, place on a cooling tray, still in a warm place, and leave them another few hours. This allows moisture to evaporate more quickly from both surfaces. Serve with a dip (see page 237). To make croutons cut your thinly rolled dough into squares or smaller triangles.

SUNFLOWER WAFERS

Sunflower seeds are one of the most nourishing foods available, an excellent source of essential fatty acids and protein. When you consider that from one tiny seed springs a magnificent plant over twelve feet tall, whose heavy flower head faces the sun and absorbs its rays all day, it is little wonder that the seeds are so energy-packed. Sunflower Wafers can be made sweet or savoury and eaten with desserts or dips.

1 C sunflower seeds, ¼C raisins or 2t tamari, a little water

Grind the seeds as finely as possible and process with the raisins or tamari. Add just enough water to make a thick dough. Nip off small pieces and press them into round flat wafers, about 1½ in/4 cm in diameter. Place on a cooling tray covered with a tea towel and leave in a warm place for several hours until they are dry and crisp.

SWEET TREATS

These munchy snacks can replace those mid-day chocolate bars or biscuits that go well with a non-coffee break. They are particularly good for children (babies too) and can quickly take the place of 'sweets'. They are also helpful as an energy 'pick-me-up' and sustainer between meals.

Most of our sweet treat ideas began one Easter when we decided that it ought to be possible to make Easter eggs that were good for you and just as palate tickling as the chocolate ones! We experimented with lots of different goodies combined in various ways and came up with some delicious 'eggs'. In fact our Easter eggs turned out more like Easter sausages but they tasted great. We often box our sweet treats attractively and give them as little gifts.

EASTER EGGS – BASIC RECIPE

1 1/2C nuts (a mixture such as pecans and hazels, or almonds and brazils), 3/4C sunflower seeds, 1/4C sesame seeds, 1C mixed dried fruit (apricots, peaches, pineapple and bananas, or pears with raisins, figs and dates), 3T dried coconut, 2T honey, juice of 1/2 orange or 1T apple or grape juice, carob powder or sesame seeds to coat the eggs

Finely grind the sesame seeds in the food processor. Add the sunflower seeds and nuts and grind well. Roughly chop the dried fruit and process with the other ingredients. Add the coconut, honey and a little fruit juice and process once more. You should end up with a slightly sticky homogeneous wodge. Powder a board with carob powder or sesame seeds and form the mixture into little balls or sausages, rolling it in the powder or seeds. Chill and keep in the fridge.

VARIATIONS AND TIPS

If you keep a few of the nuts and raisins separate, chop them coarsely and then combine them with the other ingredients, you get chewier, crunchier treats! For really nutritious treats use sunflower seeds sprouted for one or two days instead of dry ones, and roll in wheatgerm.

Liquorice Treats
For a liquorice flavour, crush a teaspoon of aniseed and mix with two tablespoons of plain or toasted sesame seeds. Use this combination to roll the treats in.

Spiced Treats
Add a pinch of allspice, cinnamon, ginger or powdered cardamom to the basic mixture and combine well. A few drops of vanilla essence are also nice.

Scented Treats
Add one or two tablespoons of orange flower or rose water to the basic mixture instead of fruit juice, or add a little finely grated orange or lemon rind.

Fruitier Treats
Try replacing half the dried fruit in the basic recipe with fresh fruit such as banana, pineapple, strawberries or cherries. You will not need to add any juice.

REFRIGERATED COOKIES

We make these with peanut butter. In fact we invented them specially for 18-year-old Jesse – the tallest member of our family – who has a passion for peanut butter. Most peanut butter you buy is actually 'cooked', but you can find raw if you look hard, or you can make your own.

1C rolled oats, ¼–½C blanched almonds, 3T peanut butter or ⅓C ground peanuts, 1–2T honey or molasses, handful of raisins, handful of dates, 1t vanilla essence, 1t cinnamon, pinch of allspice

Grind the almonds, raisins and dates in the food processor. Add the peanut butter, honey, vanilla and spices and combine well. Mix the oats with the rest of the ingredients. Form the mixture into flat cookie shapes in the palms of your hands (you may need to add a few drops of water) and place on a baking sheet. Refrigerate until firm.

APPLE SPICE BALLS

These do not keep as well as some of the other treats because of the fresh apple. Nevertheless they taste wonderful.

3–4 apples, ¹/₂C raisins, ¹/₂C coconut, ¹/₄C each of sesame, sunflower and pumpkin seeds, ¹/₄t nutmeg, ¹/₂t freshly ground cloves, 1t cinnamon, dried coconut

Finely grind the seeds and place in a bowl. Core the apples and process to a sauce, adding the spices. Stir the raisins, coconut and apple sauce into the ground seeds, form the mixture into balls, and sprinkle with some coconut. Chill and serve.

CAROB FUDGE

Once chilled these wonderful fudge balls even have the texture of ordinary fudge, and their carob flavour makes them ideal chocolate substitutes.

1C sesame seeds, ¹/₂C dried coconut, ¹/₂C carob powder, 1t honey, ¹/₂t vanilla essence

Grind the seeds very finely in the food processor. Add the other ingredients and process again. Form the mixture into little balls and chill.

INDIAN BALLS

These have an unusual and interesting spicy taste.

10 dried figs (clip off the tough stems), 6 dried dates, 1C dried coconut, 1T honey, 1T grated orange peel, 1t powdered cardamom

Grind the dried fruit finely and add the coconut, honey and orange peel. Crush several cardamom pods, remove the seeds and crush to a powder. Add to the rest of the ingredients and combine well. Form into tiny balls and refrigerate.

HALVAH

This is a favourite Middle Eastern sweet of course and one of the simplest to make.

1 C sesame seeds, 2 T honey, 2 T raw sugar, 2 T chopped nuts (pine nuts, unsalted pistachios, cashews) or 2 T raisins or 2 T finely grated carrot

Grind the sesame seeds as finely as possible. It will take a few minutes as the food processor blade has to break through the tough seed coats. Stir in the sugar, and chopped nuts (or raisins or carrot). Knead the honey into the mixture until it has a hard dough-like consistency. Form into a mini square loaf shape and chill. Serve in slices.

SUNFLOWER SNACKS

1/2 C sunflower seeds, 1/2 C carob powder, 1/4 t cinnamon, a little apple juice

Finely grind the sunflower seeds and mix with the carob and cinnamon. Add a few drops of apple juice, just enough to make the mixture bind. Form into a roll about 1 in/2.5 cm thick, chill, then slice. Alternatively, break off little bits and press them into coin-size wafers, and chill.

DRINKS

One of the greatest health secrets is to drink plenty of liquids throughout the day. This replaces vital body fluids which are constantly lost through perspiration and breathing, and helps to flush out your system. A glass of fresh vegetable or fruit juice is also a tonic, an energy booster. On an empty stomach the natural sugars, vitamins and minerals are absorbed into your bloodstream in a matter of minutes and leave you feeling clear-headed, refreshed and energised. The drinks in this section range from fruity summer thirst quenchers to protein-rich breakfast drinks.

Fruit Drinks

Fruit drinks can be made simply by juicing any kind of fresh fruit – apples, oranges, grapes, pineapples – in a centrifuge or citrus press, or by blending them with a little water in the blender. Often it is worth buying a crate of oranges or apples from a wholesaler specifically for juicing. Some of the recipes which follow call for a certain amount of, say, apple juice. You can use pure packaged juice for convenience, but fresh juice is obviously better.

FRUIT SMOOTHIES

These are made by blending a combination of fruits with a little apple juice, spring water or natural carbonated water and adding 1t honey (optional). You will need about a cupful of chopped fruit (peeled or not, depending on the fruit) and just under a cupful of liquid per person. Here are some nice combinations, which are even more scrumptious if you chill the fruit in the freezer for an hour or so first.

Banana and Peach
Banana and Strawberry/Raspberry/Blackberry/Blueberry
Banana and Apple

Pear and Apple
Mango and Orange
Peach or Apricot
Orange with a pinch of ground ginger

PINEAPPLE OR ORANGE SHAKE

For each person you will need:

½C pineapple or *orange, ¼C pineapple* or *orange juice, squeeze of lemon juice, 2 ice cubes*

Peel and pip the fruit, cut it into chunks and blend with the juice and a squeeze of lemon until smooth. Add the ice cubes and process until crushed. Serve with a sprig of mint and a twist of orange or slice of pineapple in a tall glass.

DRIED FRUIT SHAKE

For each person you will need:

⅓C dried fruit (apricots, peaches, pears, pitted prunes or dates, or a mixture), 1–2C warm water, squeeze of lemon (optional), honey (optional)

Soak the dried fruit overnight in the warm water until plump. Pour the fruit and soak water into the blender and process until smooth, adding a little lemon juice and honey if desired. A pinch of cinnamon or a drop or two of vanilla essence also go very well with the mellow flavours of dried fruit.

GARDEN PUNCH

This is our favourite summer drink. The recipe comes from a family friend who has a most beautiful garden full of flowers, fruit, vegetables and honey bees. Drinking her version of the drink on a hot summer's day surrounded by her wonderful flowers is pure paradise! To make a large jugful to quench about four thirsts you will need:

2–3C apple juice, 1C pineapple or orange juice, a handful of raspberries or blackcurrants (mainly for their colour effect!), 1 orange, 1 lemon, assorted fresh mint, fresh lemon balm, 1C fresh or dried elderflowers (de-stalked), honey, ice, 2C water

Blend the fresh mint and lemon balm with the water and berries until the leaves are finely chopped. Add the grated rinds of the orange and lemon and leave the mixture to soak in the blender for about 15 minutes. Strain into a jug and discard the leaves, berry pulp and rinds. Pour in the other juices (apple and pineapple or orange). Juice the lemon and slice the orange and add them to the jug, then add the elderflower heads (these can be strained off later, but a few poured into the glasses with the drink are particularly attractive). Sweeten with a little honey and chill. Serve in tall glasses with ice and fresh mint. As a variation try replacing the apple juice with grape juice.

Unblended Fruit Drinks

SPICED APPLE JUICE – SERVES 4

A very simple way of doing something special with plain juice.

4–6C apple juice (or organic cider!), 2T honey, 1 lemon and/or 1 orange, 6 cloves, 4 cinnamon sticks, 2 cardamom pods, pinch of nutmeg, allspice and ground cinnamon

Cut the lemon and orange into thin slices and place in a jug. Pour the apple juice over them and add the honey and spices (but not the cinnamon sticks). Cover and leave to stand for an hour. Strain the juice and serve in glasses with a stick of cinnamon in each.

LEMONADE – SERVES 4

Lemons, once used to keep clothes moths out of cupboards before moth balls were invented, have many known health benefits. Their high hesperidin content helps to strengthen collagen in the skin and blood vessels and their vitamin C soothes sore throats. Lemon juice is also supposed to be an excellent cure for hiccoughs!

3–4 lemons, 4–6C water, 1C raisins, 1T honey, 4 lemon slices

Grate the rind off one of the lemons and put it in a saucepan with about two cupfuls of water and heat to just below boiling. Strain this liquid into a bowl and add the raisins. Leave them to soak until they are plump. Pour the raisins and soak water into the blender and add the juice of the lemons, including the rindless one. Blend well and add honey to taste. Serve in tall glasses with crushed ice and a slice of lemon.

Nut and Seed Milks

These are simple to make, highly nutritious and easy to digest. They can also replace cow's milk in certain dishes. Nut milks and seed milks can be made separately or together – the principle is the same.

ALMOND MILK

This is an ambrosial introduction to these milks, and it is our favourite. We remove the almond skins as they are rather bitter and contain a high quantity of prussic acid which should be avoided. Some people blanch the almonds first, but we find it easiest to prepare the milk with unskinned almonds and then strain it through a fine sieve or piece of cheesecloth to remove the skins and pulp. As a general rule you need 1 part nuts to 3 parts water. The quantities below serve two people.

1–1½C almonds, 4C water, honey to sweeten, dash of cinnamon or nutmeg, vanilla essence (optional)

Combine almonds and water in your blender and process really well for a minute or so until the mixture is very smooth. Add the honey, cinnamon or nutmeg and vanilla.

Strain and serve. As a variation, blend a ripe banana with the almond milk.

You can use other nuts instead of almonds. Cashews are particularly good, but you may find you need a little more water. Nuts and sunflower or sesame seeds also make a nice drink.

SWEET SEED MILK – SERVES 4

1C sunflower and sesame seeds (3 parts sunflower to 1 part sesame), 4–5C water, 10 dried dates or 8 dried figs (minus the hard stalks), squeeze of lemon (optional)

Grind the seeds very finely in the blender with some of the water. When smooth, add the dates or figs and process again. The figs give a pleasing crunchy texture because of their seeds. Add the remaining water and a squeeze of lemon juice. Serve immediately. Try making Sweet Seed Milk as a breakfast drink, soaking the seeds and dried fruit together in the blender overnight. Next morning blend all the ingredients together well. The extra soaking makes the seeds and fruit even tastier and more digestible. You can use raisins or apricots instead of the dates.

VARIATIONS ON SEED AND NUT MILKS

Seed and nut milks can be flavoured in many different ways.

Banana Milk
Add two ripe bananas to the basic recipe for four people. You will need to add a little extra water, especially if you want to drink it through a straw! Blend the bananas with the milk until creamy.

Carob Milk
Add half a cupful of carob powder to the basic recipe for four people, plus a teaspoon of vanilla essence and a little extra water. Blend all the ingredients together well.

Coconut Milk

Add half a cupful of dried coconut to the basic recipe for four. Process it and the water in the blender until completely smooth, and then add the seeds or nuts and blend thoroughly.

Milk and Yoghurt Drinks

Milk and yoghurt can be used almost interchangeably to make the drinks which follow. Obviously yoghurt will give a slightly thicker, sharper drink. We recommend goat's milk for its digestibility, but cow's or even soya milk can be used too.

MOCCA MILK – SERVES 1

1 C goat's milk, 1/3C carob powder, 1t instant cereal 'coffee', 1 T honey, vanilla essence, whipped cream and chopped pecans to top (optional)

Mix the carob and a little of the milk into a paste and put it in the blender with the rest of the milk, the 'coffee' and the honey. Blend well and pour into a glass. Top with a little whipped cream and chopped pecans if desired.

BANANA MILK – SERVES 1

There are over a hundred different varieties of banana in the world, all subtly different in taste, size and colour. The bananas we know best should be eaten when their skin is speckled brown. At this stage most of the starch in them has been converted into fruit sugar, making them tastier and more digestible. Simply blend a banana and a cupful of milk in the processor and add a dash of vanilla essence. Sprinkle a few sesame seeds on the top if desired. For a thicker milk shake use a frozen banana (peel it before you freeze it) and blend as normal.

BERRY WHIP

This can be made with yoghurt or milk. If using milk, it is nice to use berries that have been frozen as they give a thicker consistency.

1 C yoghurt or milk, handful of berries (strawberries, raspberries, red- or blackcurrants), wheatgerm, honey

Blend the yoghurt or milk with the berries until smooth. If you are using thick yoghurt you may need to add a little milk to thin it. Pour into a glass and sprinkle with a little wheatgerm if desired. Drizzle with honey from a spoon and garnish with one or two whole berries.

ENERGY SHAKE

This is particularly good for athletes or dancers and can be drunk instead of breakfast. It is quick and easy to make. The blending process thoroughly breaks down the food so that it is quickly digested and assimilated and the protein and B vitamins in the recipe help to give sustained energy. The quantities given are sufficient for two servings, or one big breakfast serving.

2 C goat's milk yoghurt or nut milk, 2 T molasses or honey, 2 T wheatgerm, 2 T lecithin (optional), 1 egg yolk, 1 T tahini or finely ground sesame seeds, 2–3 drops vanilla essence.

Put all the ingredients in the blender and process thoroughly. You can add a banana or a little fresh soft fruit for flavour and fructose (quick energy). Experiment to suit your requirements and produce a drink which is just right for you.

Vegetable Juices

The amazing results of using raw juices in the treatment of cancer, colitis, diabetes, multiple sclerosis, toxemia, arthritis and rheumatism are proof of their wonderful health-giving properties. Juices are not only effective because of the

VEGETABLE JUICE CHART

Vegetable	Benefits	Notes
Beetroot	One of the best juices for helping to build up the red blood cell count in anaemia and improve the blood generally. Particularly beneficial to women suffering from menstrual disturbances. The juice from beet leaves has been found to have oestrogenic or hormonal properties and is used to increase fertility and to help women through the menopause. The minerals in beetroot juice make it a splendid liver, kidney and gall bladder cleanser.	Too strong to be drunk on its own except in very small quantities, but very palatable with other juices such as carrot. Don't forget to juice the green tops too!
Cabbage	A wonderful cleanser of the mucous membranes of the intestinal tract, cabbage juice is particularly good for treating stomach ulcers and constipation. Also good for clearing up gum infections. The juice tends to cause gas because it breaks down putrefactive matter in the intestines (a good thing).	A strong tasting juice which can be drunk on its own, but is more pleasant combined with other juices.

Carrot

The benefits of carrot juice are incredible, ranging from a digestive aid, endocrine tonic, skin cleanser and eye conditioner to a solvent for ulcerous and even cancerous growths. It is helpful in fighting infection and calming the nervous system while promoting vitality and a feeling of wellbeing. We recommend carrot as an introduction to vegetable juices.

Carrot juice can be taken in large quantities. It contains many vitamins including large amounts of vitamin A. It makes the ideal 'mixer' for other vegetable juices.

Celery

This juice is important for clearing away the dead wastes which build up in and clog the tissues of the body, causing such conditions as arthritis, diabetes, coronary disease, varicose veins, kidney stones, etc. Celery is particularly high in organic sodium, the companion to organic potassium and essential for maintaining the correct consistency of body fluids. It also contains organic calcium and other minerals which help to restore a balanced nervous system.

Good in combination with carrot juice. Together the two make a particularly well-balanced mineral drink. Juice the leafy tops too.

VEGETABLE JUICE CHART (cont.)

Vegetable	Benefits	Notes
Cucumber	A superb natural diuretic, cucumber also promotes hair growth due to its high silicon and sulphur content. It is a valuable blood pressure regulator due to its generous quantities of potassium, and also alleviates rheumatism by flushing uric acid out of the system.	This juice has a fast action and is ideal for washing the system clean, particularly several hours after eating an over-salty meal. Good with carrot and beetroot.
Lettuce	All kinds of lettuce can be juiced, and for that matter all kinds of greens. Most are very high in iron which is good as a blood builder. Also, greens are rich in chlorophyll which is exceptionally good for health. Lettuce itself is a natural tranquilliser, good for soothing the stomach and a gentle diuretic.	It is best to juice the outer leaves which are highest in chlorophyll. Often they are less tender and therefore not as nice for salads. Lettuce is good combined with other vegetables.
Parsley	Although a herb, parsley juice taken in small amounts (1 or 2T) with other juices is highly beneficial. It is good for the adrenal and thyroid glands, for maintaining a healthy genito-urinary system and helping kidney problems, and for treating eye disorders.	Should be taken in small amounts as it is very potent. Best combined with carrot or carrot and celery.

Spinach	Spinach has probably the best action of any vegetable juice on the entire digestive system. It not only cleanses and cures constipation, but helps to heal the lining of the entire tract, in particular the colon and small intestine. Its action is primarily due to oxalic acid. Unlike the inorganic oxalic acid found in cooked spinach which forms harmful crystals, organic oxalic acid stimulates peristalsis (wave-like contractions of the muscles in the gut). Thus spinach juice helps to speed up digestion, creating a fast transit time for food and wastes.	Spinach juice can be rather strong on its own and is therefore best combined. Other greens which contain organic oxalic acid are: Swiss chard, beet greens, kale, turnip greens, broad-leafed French sorrel.
Watercress	Exceptionally high in sulphur, watercress juice is a good intestinal cleanser and is helpful, in combination with other juices, for anaemia, haemorrhoids and emphysema.	Should always be combined with other vegetables, and very little should be used as the juice is extremely bitter. Add a little to carrot and spinach juice.

vital enzymes, minerals and vitamins they contain, but because these nutrients are so readily available. Juices are absorbed and assimilated very rapidly, sometimes in a matter of minutes. This is why they make such good pick-me-up drinks. Try a glass of carrot and apple juice after a sleepless night and see how it galvanises your whole system and sets you on your feet ready to face the demands of the day.

When you make juice, be sure to put ice in the container you pour it into. This helps to reduce oxidation. Although it is best drunk immediately, juice can be kept for several hours in an airtight jar in the fridge or in a thermos flask with ice.

As the chart on pages 278–81 shows, each vegetable has specific properties which are helpful for different ailments. The benefits and tastes of various vegetable juices complement each other. We think the best tasting juice is carrot and apple (5 parts carrot to 3 parts apple). It should encourage you to try some of the other combinations such as:

Carrot, beet and cucumber (8:2:3)
Carrot, celery and spinach (7:5:4)
Carrot, cabbage and lettuce (8:4:4)

A little parsley or watercress and a few fresh herbs, such as mint, chives or basil, can be added to any of these. Also one or two fresh tomatoes. If you blend in a handful of sunflower seeds or blanched almonds you get a delicious and sustaining protein drink. Extra sprouts, alfalfa especially, are also good juiced along with other vegetables.

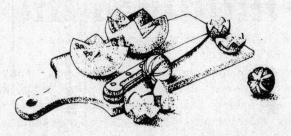

Teas and Tisanes

Although not raw, tisanes or herbal teas are so beneficial as body cleansers and strengtheners that we feel they should be part of a raw diet. Another good reason for including them is that they do not have the irritant effects of tea and coffee. Herb teas have a wide range of medicinal properties which have been used to cure all sorts of ailments over the centuries from rheumatism (agrimony) to 'those much given to sighing' (marjoram). As you will see there is some overlap between these groups. Often one herb is helpful for several conditions.

For indigestion/stomach ache – *peppermint, caraway, dill, fennel, aniseed, lemon grass, lemon balm, sweet cicely*

Diuretics for weight loss/for the kidneys – *celery seed, marsh mallow, dandelion, couch grass, golden rod, agrimony*

For the liver – *agrimony, mugwort, angelica*

For infections/colds – *rose hip, coltsfoot, comfrey, aniseed, horehound, liquorice, sage*

Diaphoretic for producing a sweat and reducing a fever – *lime/linden, peppermint, elderflower, yarrow*

Sleep inducing/calming the nerves – *chamomile, hops, lime, skullcap, orange blossom, passion flower, red clover*

As a tonic – *nettle, mint, ginseng, rosemary, blackberry, raspberry and strawberry leaf*

Skin disorders – *celandine, golden rod, St John's Wort*

Some of the most commonly found and best-tasting herb teas are: *peppermint, rose hip, jasmine, orange blossom, chamomile, fennel, lime or linden blossom, golden rod, lemon verbena, hibiscus, and lemon grass.* Although they all have some health-benefiting properties they are also drunk purely for their flavour. Delicious mixtures of herb teas can be found in health food shops. Those that come in teabag form are very convenient, but expensive. Some health food shops make up their own herb tea mixtures which are less pricey. You can do the same. Just buy small quantities of different herbs and start experimenting.

TO MAKE HERB TEA

You will need about a tablespoon of dried herbs (one sort only or a mixture) to make two cups. Steep the herbs for 5-10 minutes, stirring occasionally to extract the full flavours. Then strain the tea and serve with a slice of lemon or lime and a little honey, if desired. Some people like to add a spoonful of fresh cream or a little milk.

Herb teas are also wonderful iced. In the summer we like to keep a teapot full of our favourite tea in the fridge, often drinking it in place of snacks. Iced tea needs to be stronger than hot. If you want honey, though, stir it in while the tea is still hot. Refrigerate the tea until it is needed and then serve in tall glasses with ice and a twist of lemon, or a pair of cherries sitting on the edge. It is also nice to freeze small flowers such as honeysuckle, lilac or elderflower into cubes of ice and float these in the tea! Some herbal flowers such as hibiscus give such a beautiful red colour to teas that it is a crime not to serve them in a glass.

COOKED FOODS – the other 25 per cent

Cooked foods are never the focus of any of our meals. Pride of place is always given to salads, croquettes and so on. However, here are some of the cooked foods we include in our 75 per cent raw regime.

Fish   Lightly grilled with lemon juice and herbs. We very occasionally eat fish raw, Japanese style, with lemon juice and finely grated white radish, turnip or horseradish.

Lamb's liver   An excellent source of vitamins and good for growing bodies. We fry small slivers in a little olive oil with a few onion rings and chopped mushrooms.

Game/free-range poultry   Simply baked in a slow oven with herbs, and sometimes with an oatmeal, onion and herb stuffing to soak up the cooking juices.

Fresh seafood   (lobster, crayfish, crab, prawns, scallops, oysters) Wonderful for trace elements. We poach them gently, if necessary, in as little water as possible, and serve them in a splendid *fruits de mer* salad.

Laver bread   A surprisingly tasty dish of seaweed mixed with rolled oats and chopped spring onions, seasoned with tamari, made into patties and quickly fried in the minimum of olive oil.

Free-range eggs   Occasionally we blend the raw yolks into protein and energy drinks, otherwise we boil them for the minimum length of time and chop them into salads.

Potatoes   Baked is best. To make potato salad we put the potatoes, well scrubbed, into the oven for about 40 minutes; remove them, scrape the flesh out and mix with chives and a yoghurt dressing. We also fill the shells with finely chopped salad or melted cheese.

Whole brown rice, millet or buckwheat   Boiled in just enough water and no more. We mix these with salads or stir-fry them with a few vegetables.

Toasted goodies (nuts, seeds, sprouts)   We probably do more toasting than any other sort of cooking! Toasting makes things crunchy and deliciously different from the raw article. Nuts and seeds we usually put in a thin layer in a shallow oven-proof dish and toast under the grill for a minute or two until they are golden brown. We stir them around to make sure they brown on all sides. Pumpkin, sunflower and sesame seeds actually pop like corn and can be sprinkled with a little garlic powder or vegetable bouillon for added flavour. Wheat, barley and bean sprouts do better on a baking tray in a moderate oven; they take about 15–20 minutes to go crispy. We use them as croutons or grind them to a fine powder to sprinkle on milk drinks.

Roasted/fried spices The full flavour of spices can be released by dry roasting or frying in oil. Simply heat them dry or in a little oil in a small saucepan for a couple of minutes until they change colour and you can smell their oils being released. Remove from the heat, crush and use immediately. Caraway, coriander, poppy and cumin seeds are especially good fried or dry roasted.

Soups Especially welcome on a bleak winter's day. Whether meat- or vegetable-based you can at least be sure of getting most of the minerals, if not the vitamins, in the broth.

REFERENCES

REFERENCES

Chapter 1

Avoiding Heart Attacks, HMSO, Department of Health and Social Services, London, 1981

Bircher, R. ed. *Way to Positive Health*, Bircher-Benner Verlag, Erlenbach-Zurich, 1967

Bircher, R. 'A Turning Point in Nutritional Science', reprint from *Lee Foundation for Nutritional Research*, No. 80, Milwaukee, Wisconsin

Colgan, M. *Your Personal Vitamin Profile*, Quill; New York, 1982

Compendium of Health Statistics, 4th ed., Office of Health Economics, London, 1981

Ehret, Arnold. *Mucusless Diet Healing System*, Ehret Literature Publishing Co., Cody, Wyoming, 1953

First Health and Nutrition Examination Survey, United States 1971–1972, DHEW publication No. 76–1219–1, Rockville, Maryland, 1976

Gerson, Max. *A Cancer Therapy: Results of 50 Cases*, Totality Books, Del Mar, California, 1977

Healthy People: The Surgeon General's Report on Health Promotion and Disease Prevention, DHEW, publication Nos. 79-55071 and 79-55071A, Washington, DC, 1979

Kollath, W. 'Der Vollwert der Nahrung und seine Bedeutung für Wachstum und Zellersatz', *Experimentelle Grundlagen*, Stuttgart, 1950

Levy, R. I. Address to the National Institute of Health, 27 November 1979

National Cancer Institute Monograph 57, National Cancer Institute publication No. 81-2330, Bethesda, Maryland, June 1981

Nolfi, Kristine. *My Experiences with Living Foods*, Humlegaarden, Humlebaek, Denmark (undated)

Pelletier, Kenneth R. *Holistic Medicine – from Stress to Optimum Health*, Delacorte Press/Seymour Lawrence, New York, 1979

Pottenger, F. M. Jr. 'The Effect of Heat Processed Foods', *American Journal of Orthodontistry and Oral Surgery*, Vol. 32, No. 8, pp.467–85, 1946

Pottenger, F. M. Jr. *Pottenger's Cats*, Price-Pottenger Nutrition Foundation, La Mesa, California, 1983

Pottenger, F. M. Jr., and Simonsen, D. G. 'Heat Labile Factors Necessary for the Proper Growth and Development of Cats', *Journal of Laboratory and Clinical Medicine*, Vol. 25, p.6, 1939

Price, Weston A. *Nutrition and Physical Degeneration*, Price-Pottenger Nutritional Foundation, La Mesa, California, 1970

Straus, Charlotte Gerson. A lecture presented before the Second Annual Cancer Convention of the Cancer Control Society, Los Angeles, California, 1974

Walker, Norman W. *Natural Weight Control*, O'Sullivan Woodside & Co., Phoenix, Arizona, 1981

Williams, R. J. and Kalita, Dwight K. eds. *A Physician's Handbook on Orthomolecular Medicine*, Keats Publishing Co., New Canaan, Connecticut, 1977

Chapter 2

Bicknell, F., and Prescott, F. *The Vitamins in Medicine*, Lee Foundation for Nutritional Research, Milwaukee, Wisconsin, 1953

Bircher, Ralph. *A Turning Point in Nutritional Science*, Lee Foundation for Nutritional Research, Milwaukee, Wisconsin (undated) (also personal communication with author)

Culverwell, M. 'Cooking Methods Linked with Cancer', in *University of California Clip Sheet*, 29 December 1981

Douglass, J. M. 'Is "How Vitamin C Works" Medicine's Best Discovery?', *Nutrition Today*, March/April 1980 (also personal communication)

Evans, R. J. and Butts, H. A. 'Inactivation of Amino Acids by Autoclaving', *Science*, Vol. 109, p.569, 1949

Gerber, Donald. *Medical World News*, 13 February 1970

Glatzel, H. *Verhaltensphysiologie der Ernährung*, Berlin, 1973

Greiser, Norburt. 'Zur Biochemie der Vorspeise', *Deutsches Medizinisches Journal*, 15/8, 20/4, pp.267–71, 1964

Hein, R. E. and Hutchings, I. J. *Nutrients in Processed Foods*, ed. American Medical Association on Foods and Nutrition, Publishing Sciences Group, Acton, Massachusetts, 1974

Kollath, Werner. *Die Ordnung Unserer Nahrung*, Hippocrates-Verlag, Stuttgart, 1955

Kouchakoff, Paul. 'The Influence of Food Cooking on the Blood Formula of Man', *Proceedings of First International Congress of Microbiology*, Paris, 1930

Kuratsune, Masanore. 'Experiment of Low Nutrition with Raw Vegetables', in *Kyushu Memoirs of Medical Science*, Vol. 2, No. 1–2, June 1951 (also in *Der Wendepunkt*, pp.226–7, 1965)

Pottenger, F. M. Jr. *Pottenger's Cats*, Price-Pottenger Nutrition Foundation, La Mesa, California, 1983

Priestley, R. J. ed. *Effects of Heating on Foodstuffs*, Applied Science Publishers, London 1979

Schroeder, H. A. Article in *American Journal of Clinical Nutrition*, Vol. 24, 1971

Senti, F. R. *Nutrients in Processed Foods*, ed. American Medical Association Council on Foods and Nutrition, Publishing Sciences Group, Acton, Massachusetts, date unknown

Walsh, Michael J. Article in *Modern Nutrition*, March 1968

Yamaguchi, T. et al. Article in *Journal of Vitaminology*, Vol. 5, p.88, 1959

Chapter 3

Aly, Karl-Otto. 'Cancer defeated by the body's own defenses', *Tjidskrift fur Halso*, 9 September 1965

Anderson, James W. *Diabetes*, Martin Dunitz, London, 1981

Bircher, R. ed. *Way to Positive Health*, Bircher-Benner Verlag, Zurich, 1967

Bircher-Benner, M. *The Prevention of Incurable Disease*, James Clarke & Co., Cambridge, 1981

Brekhman, I. I. *Man and Biologically Active Substances*, Pergamon Press, Oxford, 1980

Cheraskin, E., Ringsdorf, W. M., and Clark, J. W. *Diet and Disease*, Rodale Press, Emmaus, PA, 1968

Douglass, J. M. 'Raw Diet and Insulin Requirement', *Annals of Internal Medicine*, Vol.82, p.61, 1975

Douglass, J. M. and Douglass, S. N. 'Sugar, Cereals and Sharks', *New England Journal of Medicine*, Vol.295, p.788, 1976

Douglass, J. M. et al. 'Nutrition, non-thermally prepared Food and Nature's Message to Man', *Journal of the International Academy of Preventative Medicine*, Vol. VII, No. 2, July 1982

Douglass, J. M. and Rasgon, I. 'Diet and Diabetes', *Lancet*, December 1976

Gerson, Max. *A Cancer Therapy*, Totality Books, Del Mar, California, 1977

Goldin, Barry, et al. 'Effect of Diet and Lactobacillus acidophilus Supplements on Human Fecal Bacterial Enzymes', *Journal of Natural Cancer Institute*, Vol. 64, No. 2, February 1980

Hare, D. C. 'A Therapeutic Trial of a Raw Vegetable Diet in Chronic Rheumatoid Conditions', *Proceedings of The Royal Society of Medicine*, Vol. XXX, 1, 13 October, 1936

Kelley, W. D. *One Answer to Cancer*, The Kelley Research Foundation, 1971

Kuhl, Johannes. *Checkmate for Cancer*, Viadrina Verlag A. Trowitzh, Braunlage, Germany.

Liechti-von Brasch, D. '70 Jahre Erfahrungsgut der Bircher-Benner Ordnungstherapie', *Erfahrungsheilkunde*, Band XIX, Heft 6/7/8, 1970

Livingston, Virginia, and Wheeler, Owen. *Food Alive*, The Livingston Wheeler Medical Clinic, San Diego, 1977

Manner, Harold. Natural Health Federation Conference, Chicago, September 1977

Neiper, Hans. Lecture given at the Royal Society of Medicine, London, May 1980

Peto, R., Doll, R. et al. 'Can dietary beta-carotene materially reduce human cancer rates?', *Nature*, Vol. 290, 19 March 1981

Shamburger, R. J. and Willis, C. E. J. *National Cancer Institute*, Vol. 48, No. 5, pp.1491–7, 1972

Tropp, K. 'Die Pflanzenfermente der Rohkost und ihre Bedeutung für den Verdauungskanal des Menschen', *Der Wendepunkt*, May 1948

Chapter 4

Abrams, G. D. and Bishop, J. E. 'Normal Flora and Leukocyte Mobilization', *Archives of Pathology*, Vol. 70, pp.213–17, February 1965

Airola, P. *Arthritis Can Be Cured*, date unknown

Alvarez, Walter C. 'Enzyme defects can induce cell aging', *Geriatrics*, p.72, August 1970

Bircher, Ralph. *Gesünder durch weniger Eiweiss, Geheimarchiv der Ernährungslehre, Höchstleistungskost für Sport, Berg, Eis, Wüste und Dschungel, Sturmfeste Gesundheit* and *HUNSA Das Volk, das keine Krankheit kannte*. All published in the series 'Edition Wendepunkt', Bircher-Benner Verlag 1980/1.

Brekhman, I. I. and Dardymov, I. V. 'New Substances of Plant Origin which increase Non-specific Resistance', *Annual Review of Pharmacology*, Vol. 9, 1969

Cheney, G. 'Prevention of histamine-induced peptic ulcers by diet', *Stanford Medical Bulletin*, Vol. 6, p.344, 1948

Cheney, G. 'The nature of the anti-peptic ulcer dietary factor', *Stanford Medical Bulletin*, Vol. 8, p.144, 1950

Cheney, G. 'Anti-peptic ulcer dietary factor', *American Dietary Association*, vol. 26, p.9, September 1950

Dineen, P. 'Effect of alteration in intestinal flora on host resistance in systemic bacterial infection', *Infectious Diseases 109*, November/December 1961

Hill, M. J. et al. 'Bacteria and Etiology of Cancer of the Large Bowel', *The Lancet*, vol. 1, pp.95–100, 1970

Karstrom, Henning. *Protectio Vitae*, February 1972

Karstrom, Henning. *Rätt Kost*, Skandinavska Bokforlaget, Gavle, 1982

Lai, Chiu-Nan. 'The active factor in wheat sprout extract inhibiting the metabolic activation of carcinogens in vitro', *Nutrition and Cancer*, Vol. 1, No. 3, pp.9–21, date unknown

Lai, Chiu-Nan. 'Anti-mutagenic activities of common vegetables and their chlorophyll content', *Mutation Research*, Vol. 77, pp.245–50, 1980

Offenkrantz, W. G. 'Water Soluble Chlorophyll in the Treatment of Peptic Ulcers of Long Duration', *Review of Gastroenterology*, Vol. 17, p.359, 1950

Pahlow, Mannfried. *Living Medicine*, Thorsons, Wellingborough, 1980

Pearson, D. and Shaw, S. *Life Extension*, Warner Books, New York, 1982

Peterson, Vicki. *Eating Your Way to Health*, Allen Lane, London, 1981

Reddy, Bandaru. 'Effects of High Risk and Low Risk Diets for Colon Carcinogenesis of Fecal Microflora and Steroids in Man', *Journal of Nutrition*, July–December, 1975

Rose, Geoffrey. 'Colon cancer and blood cholesterol', *The Lancet*, 9 February 1974

Schmidt, Siegmund. 'Cancer and Leukemia', reprint article from Dr Schmidt, 4502 Bad Rothenfelds T.W., Bez, Osnabruck, Germany

Talbot, John. 'Role of dietary fibre in diverticular disease and colon cancer', *Federation Proceedings*, Vol. 40, No. 9, July 1981

Thornton, J. R. 'High Colonic pH Promotes Correctal Cancer', *The Lancet*, 16 May 1981

Virtanen, A. I. 'Die Enzyme in Lebendigen Zellen', *Suomen Kemistilehti*, B. XV, 1942

Virtanen, A. I. *Angewandte Chemie*, Vol. 70, No. 17–18, pp.544–52, 1958

Virtanen, A. I. *Suomen Kemistilehti*, No. 4, pp.108–25, 1964

Virtanen, A. I. Report on primary plant substances and decomposition reactions in crushed plants, Biochemical Institute, Helsinki, 1964

Chapter 5

Blauer, Stephen. *Rejuvenation*, Hippocrates Health Institute, Boston, 1980

Eastwood, M. A. and Mitchell, W. D. 'The place of vegetable fibre in the diet', *British Journal of Hospital Medicine*, p.123, January 1974

Ershoff, B. H. 'Antitoxic Effects of Plant Fiber', *American Journal of Clinical Nutrition*, Vol. 27, pp.1395–8, December 1974

Gabor, M. 'The anti-inflammatory action of flavinoids', *Akademial Kiado*, Budapest, 1972

Gelin, L. E. 'Rheologic disturbances and the use of low viscosity dextran in surgery', *Review of Surgery*, Vol. 19, pp. 385–400, 1962

Hughes, J. H. and Latner, A. L. 'Chlorophyll and haemoglobin regeneration after haemorrhage', *Journal of Physiology*, Vol. 86, p.338, 1936

Hughes, R. E. and Wilson, H. K. 'Flavinoids: some physiological and nutritional considerations', *Progress in Medical Chemistry*, Vol. 14, pp.285–300, 1977

Jean, V. *Aromathérapie*, Librairie Maloine S.A., Paris, 1974

Kelsay, J. 'A review of research on effects of fibre intake on man', *American Journal of Clinical Nutrition*, Vol. 31, pp.142–59, January 1978

Knisely, M. H. et al. 'Sludged Blood', *Science*, Vol. 106, pp.431–40, 1947

Lai, Chiu-Nan. 'The active factor in wheat sprout extract inhibiting the metabolic activation of carcinogens in vitro', *Nutrition and Cancer*, Vol. 1, No. 3, pp.19–21, date unknown

Lai, Chiu-Nan et al. 'Anti-mutagenic activities of common vegetables and their chlorophyll content', *Mutation Research*, Vol. 77, pp.245–50, 1980

Lewis, W. H. and Memory, P. F. Elvin-Lewis. *Medical Botany*, John Wiley, New York, 1977

Offenkrantz, W. G. 'Water Soluble Chlorophyll in the Treatment of Peptic Ulcers of Long Duration', *Review of Gastroenterology*, Vol. 17, p.359, 1950

Patek, A. J. 'Chlorophyll and the Regeneration of the Blood', *Annals of Internal Medicine*, Vol. 57, p.73, 1936

Pfeiffer, Carl C. *Mental and Elemental Nutrients*, Keats Publishing Co., New Canaan, Connecticut, 1975

Robbins, R. C. 'Effects of vitamin C and flavinoids on blood cell aggregation and capillary resistance', *Internationale Zeitung Vitaminforschung*, Vol. 36, pp.10–15, 1966

Robbins, R. C. 'Action of flavinoids on blood cells: trimodal action of flavinoids elucidates their inconsistent physiologic effects', *International Journal of Vitamin and Nutrition Research*, Vol. 44, pp.203–16, 1974

Robbins, R. C. 'On Bioflavinoids: new findings about a remarkable plant defence against disease and its dietary transfer to man', *Executive Health*, Vol. XVI, No. 12, 1980

Spiller, Gene. 'Interaction of dietary fiber with other dietary components: a possible factor in certain cancer etiologies', *American Journal of Clinical Nutrition*, pp.S231–2, October 1978

Wattenberg, L. W. et al. 'Induction of increased benzpyrene hydroxylase activity by flavones and related compounds', *Cancer Research*, Vol. 28, pp.934–7, 1968

Chapter 6

Becker, Robert O. 'Electromagnetic Forces and Life Processes', *Technological Review*, December 1972

Bircher, Ralph. 'A Turning Point in Nutritional Science', *Lee Foundation for Nutritional Research*, Milwaukee, Wisconsin, Reprint No. 80.

Bircher, Ralph. *Way To Positive Health and Vitality*, Bircher-Benner Verlag, Zurich, 1967

Bircher-Benner, M. *The Essential Nature and Organization of Food Energy and the Application of the Second Principle of Thermo-Dynamics to Food Value and its Active Force*, John Bale Sons & Curnow, London, 1939

Bircher-Benner, M. *The Prevention of Incurable Disease*, James Clarke, 1981

Bohm, David. *Wholeness and the Implicate Order*, Routledge & Kegan Paul, London, 1980

Bray, H. G. and White, K. *Kinetics and Thermodynamics in Biochemistry*, Churchill, 1957

Brekhman, I. I. *Man and Biologically Active Substances*, Pergamon Press, Oxford, 1980

Burr, Harold Saxton. *Blueprint for Immortality: The Electric Patterns of Life*, Neville Spearman, London, 1952

Crile, George. *The Bipolar Theory of Living Processes*, Macmillan, New York, 1926

Crile, George. *The Phenomena of Life: A Radio-Electrical Interpretation*, W. W. Norton, New York, 1936

Dakin, H. S. *High Voltage Photography*, 2nd ed., 3101 Washington St, San Francisco, California 94115, 1975

Eppinger, Hans. 'Transmineralisation und vegetarische Kost', in *Ergebnisse der Inneren Medizin und Kinderheilkunde*, Vol. 51, 1936.

Eppinger, Hans. 'Über Rohkostbehandlung', *Wiener Klinische Wochenschrift*, No. 26, pp.702–8, 1938

Eppinger, Hans. 'Die Permeabilitätspathologie als Lehre vom Krankheitsbeginn', Vienna, 1949

Galvani, Luigi. *Commentary on the Effect of Electricity on Muscular Motion – a Translation of Luigi Galvani's De Virbis Electricitatis in Motu Musculari Commentarius*, E. Licht, Cambridge, Massachusetts, 1953

Gerson, Max. *A Cancer Therapy*, Totality Books, Del Mar, California, 1977

Hauschka, Rudolph. *The Nature of Substance*, Vincent Stuart, London, 1966

Irons, V. E. *There is a Difference*, pamphlet published by V. Irons, Natick, Massachusetts 01769, on Pfeiffer's work (undated)

Karstrom, Henning. *Rätt Kost*, Skandinavska Bokforlaget, Gavle, 1982

Kuhn, Thomas. *The Structure of Scientific Revolutions*, University of Chicago Press, 1962

Lakhovsky, G. *L'Origine de la Vie*, Editions Nilsson, Paris, 1925

Lund, E. J. *Bioelectric Fields and Growth*, University of Texas Press, 1947

Moss, T. *The Probability of the Impossible*, J. P. Tarcher, Los Angeles, 1974

Pardee, A. B. and Ingraham, L. L. 'Free energy and entropy in metabolism', in *Metabolic Pathways*, Vol. 1, ed. D. M. Greenberg, date unknown

Pfeiffer, E. *Formative Forces in Crystallization*, Rudolph Steiner Publications, London, 1936

Schrödinger, Erwin. *What is Life? and Mind and Matter?*, Cambridge University Press, 1980

Seeger, P. G. *Archive experimentelle Zellforschung*, Vol. 20, p.280, 1937, Vol. 21, p.308, 1938

Simoneton, A. *Radiations des aliments ondes humaines, et santé*, Le Courrier du Livre, Paris, 1971

Szent-Györgyi, Albert et al. 'How vitamin C really works . . . or does it?' *Nutrition Today*, p.6, September/October 1979

Thompkins, P. and Bird, C. *The Secret Life of Plants*, Allen Lane, London, 1974

Warburg, Otto. Article in *Naturwissenschaften*, Vol. 21, p.485, 1954

Williams, R. 'A Renaissance of Nutritional Science is Imminent', *Perspectives in Biology and Medicine*, Vol. 17, No. 1, Autumn 1973

Zukav, G. *The Dancing Wu Li Masters*, William Morrow, New York, 1979

Chapter 7

Bailey, Covert. *Fit or Fat*, Pelham Books, London, 1980

Beinhorn, G. *Food for Fitness*, World Publications, Mountain View, California, 1975

Bircher-Benner, M. O. *Food Science for All*, C. W. Daniel, London, 1928

Bircher R. ed. *Way to Positive Health and Vitality*, Bircher-Benner Verlag, Erlenbach-Zurich, 1967

Chittenden, Russell H. *Physiological Economy in Nutrition*, Heinemann, London, 1905

Chittenden, Russell H. *The Nutrition of Men*, Stokes, New York, 1907

Eimer, K. 'Klinik Schwenkenhacher', *Zeitschrift für Ernährung*, July 1933

Goulart, F. S. *Eating to Win*, Stein and Day, New York, 1980

Mirkin, Gabe. *The Sportsmedicine Book*, Little, Brown and Co., Boston, 1978

Nutrition Today, Vol. 3, No. 2, June 1968

Physician and Sportsmedicine Magazine, January 1976

Ostrand, Per-Olaf and Rodahl, Kaare. *Textbook of Work Physiology*, McGraw-Hill, New York, 1970

Chapter 8

Bircher, Ralph. 'The Question of Protein', photocopied report from author

Bircher, Ralph. 'A Turning Point in Nutritional Science', photocopy from author, see reference in Chapter 1

Fridovich, I. 'Superoxide Dismutase: a dramatic new enzyme discovery that protects against radiation and prevents disease', *Bestways*, August 1980

Issels, Josef. 'Nutritional Protection against Cancer', *Tjidskrift fur Halsa*, Stockholm, Nos. 1, 2, 3, 1972

Karstrom, Henning. *Rätt Kost*, Skandinavska Bokforlaget, Gavle, 1982

Katenkamp and Stiller. 'Das Amyloid', *Hippokrates*, Vol. 41, No. 1, pp.5–23, 1970

Katenkamp and Stiller. *Histochemistry of Amyloid*, VEB Gustav Fischer Verlag, Jena, 1975

Kruijswijk, H., Oomen, H. A. and Hipsley, E. H. *Voeding*, 30/5, pp.225–230, 1969

McCully, Kilmer S. *American Journal of Clinical Nutrition*, Vol. 28, May 1975 and *American Journal of Pathology*, Vol. 59, pp. 181–93, Vol. 61, pp.1–8, 1970

McCay, Clive M. 'Effects of restricted feeding upon aging, etc.', *American Journal of Public Health*, Vol. 37, May 1947

McCay, Clive M. *Notes on the History of Nutrition Research*, Hans Huber, Bern, 1973

Pfeiffer, C. and Banks, J. *Total Nutrition*, Simon and Schuster, New York, 1980

Rosenfield, Albert. *Pro-Longevity*, Avon, New York, 1976

Thomas, W. A. et al. *American Journal of Cardiology*, January 1960, and *American Medical Association*, new release, 21 June 1965

Walford, Roy L. *Maximum Life Span*, W. W. Norton & Co., London, 1983

Winick, M. 'Slow the Problems of Aging and Quash its Problems – With Diet', *Modern Medicine*, pp.68–74, 15 February 1978

Chapter 9

Ahrens, E. H. and Connor, W. E. *American Journal of Clinical Nutrition*, Vol. 32, 1979

Bennett, W. and Gurim, J. *The Dieter's Dilemma*, Basic Books, New York, 1982

Bierman. *Nutrition and Aging*, ed. M. Winick, John Wiley, New York, 1976

Bircher-Benner Clinic Staff. *Bircher-Benner Keep Slim Nutrition Plan*, Nash Publishing, Los Angeles, 1973

Carlson, A. J. *The Control of Hunger in Health and Disease*, University of Chicago Press, 1914

Chandra, R. K. *Federal Proceedings*, Vol. 39, p.3088, 1980

Douglass, J. 'Diet and Diabetes', *The Lancet*, 11 December 1976, also personal communication

Enzi, G. et al. *Obesity: Pathogenesis and Treatment*, Academic Press, New York, 1981

Healthy People: The Surgeon General's Report on Health Promotion and Disease Prevention, DHEW publication, Nos. 79-55071 and 79-55071A, Washington DC, 1979

Nolfi, Kristine. *My Experiences with Living Food*, Humlegaarden, Denmark (undated)

Randolph, T. G. 'Masked Food Allergy as a Factor in the Development and Persistence of Obesity', *Journal of Laboratory and Clinical Medicines*, Vol. 32, p.1547, 1947

Rinkel, H. J., Randolph, T. G. and Zeller, Michael. *Food Allergy*, Charles C. Thomas & Co., Springfield, 1951

Vahonny, G. V. *American Journal of Clinical Nutrition*, Vol. 35, p.152, 1982

Walker, Norman W. *Natural Weight Control*, O'Sullivan Woodside & Co., Phoenix, Arizona, 1981

Williams, R. *Nutrition Against Disease*, Pitman Publishing Co., New York, 1971

Chapter 10
Airola, Paavo O. *How to Keep Slim With Juice Fasting*, Health Publishers, Phoenix, Arizona, 1971

Applegate, W. V. and Connolly, P. Price-Pottenger Lectures, 1974

Botman, S. G. and Crombie, W. M. *Journal of Experimental Botany*, Vol. 9, p.52, 1958

Cancer Control Journal, September/December 1970, Los Angeles, California

Curtis, C. 'An Account of the Diseases of India as they appeared in the English Fleet', Edinburgh, 1807

Hegazi, S. M. *Zeitschrift für Ernährungswissenschaft*, Vol. 13, p.200, 1974

Kakade, M. L. and Evans, R. J. *Journal of Food Science*, Vol. 31, p.781, 1966

Kirchner, H. E. *Nature's Healing Grasses*, H. C. White Publications, Riverside, California, date unknown

Kirchner, H. E. *Live Food Juices*, H. E. Kirchner Publications, Monrovia, California, 1957

Kohler, G. O. 'Unidentified Factors Relating to Reproduction in Animals' Feedstuffs', 8 August 1953

Kulvinskas, V. *Nutritional Evaluation of Sprouts and Grasses*, Omango D' Press, Weathersfield, Connecticut, 1978

Kuppuswamy, S. 'Proteins in Food', *Indian Council of Medical Research*, New Delhi, 1958

Leichti-von Brasch, D. et al. *Bircher-Benner Keep Slim Nutrition Plan*, Nash Publishing, Los Angeles, 1973

Mayer, A. M. and Poljakoff-Mayber, A. *The Germination of Seeds*, Pergamon Press, Oxford 1966

Price, W. *Nutrition and Physical Degeneration*, Price-Pottenger Foundation, La Mesa, California, 1970

Price-Pottenger Foundation. *The Guide to Living Foods Workbook*, La Mesa, California, 1978

Sellman, Per and Gita. *The Complete Sprouting Book*, Turnstone Press, Wellingborough, England, 1981

Szekely, E. B. *The Essene Gospel of Peace*, International Biogenic Society, Cartago, Costa Rica, 1978

Tsai, C. Y., Dalby, A. et al. 'Lysine and Tryptophan Increases During Germination of Maize Seed', *Cereal Chemistry*, Vol. 52, No. 3, 1975

Walker, N. W. *Raw Vegetable Juices*, Jove/Harcourt Brace Jovanovich, New York, 1977

Walker, N. W. *Natural Weight Control*, O'Sullivan Woodside & Co., Phoenix, Arizona, 1981

Chapter 11

Almond, S. and Logan, R. F. L. 'Carotenemia', *British Medical Journal*, Vol. 2, pp.239–41, 1942

Bernstein, D. S. and Wachman, A. 'Diet and Osteoporosis', *The Lancet*, Vol. 7549, p.958, 1968

Clemetson, C. A. B. 'Bioflavinoids as Antioxidants for Ascorbic Acid', Symposium sui Bioflavinoidi, Stresa, 23–25 April 1966

Clemetson, C. A. B. et al. 'Estrogens in Food: The Almond Mystery', *International Journal of Gynaecology and Obstetrics*, Vol. 15, pp.515–21, 1978

Clemetson, C. A. B. and Andersen, L. 'Plant Polyphenols as antioxidants for ascorbic acid', *Annals of the New York Academy of Sciences*, Vol. 136, Art. 14, pp.339–78, date unknown

Curry, S. B. 'Aspects morpho-histochimiques et biochimiques du tissu adipeux dans la dermohypodermose cellulitique', paper given at Cème Journées de Médicine Esthétique, Monte Carlo, 14–15 May 1978

Dukan, Pierre. *La Cellulite en Question*. La Table Ronde Eds., Paris

First Health and Nutrition Examination Survey, United States 1971–1972, DHEW publication, 76-1219-1 Rockville, Maryland, 1976

Kemmann, E. et al. 'Amenorrhea associated with carotenemia,' *Journal of the American Medical Association*, Vol. 249, No. 7, pp.9206–929, 18 February 1983

Pfeiffer, Carl C. *Mental and Elemental Nutrients*, Keats Publishing Co., New Canaan, Connecticut, 1975

Schroeder, H. A. *American Journal of Clinical Nutrition*, Vol. 24, p.562, 1971

Taska, Richard Jr. *Life in the Twentieth Century*, Omangod Press, Woodstock Valley, Connecticut, 1981

Chapter 12
Armstrong, Bruce. *American Journal of Clinical Nutrition*, December 1979
Bircher, Ruth. *Eating Your Way to Health*, Faber and Faber, London, 1961
Bircher, R. ed. *Way to Positive Health and Vitality*, Verlag Bircher-Benner, Erlenbach-Zurich, 1967
Douglass, John. 'Nutrition, Nonthermally-Prepared Food and Nature's Message to Man', *Journal of the International Academy of Preventative Medicine*, Vol. VII, No. 2, July 1982, also private communication with author
Formica, Palma. Article in *Current Therapeutic Research*, March 1962
Meneely, George R. *Nutrition Review*, 1976
Pfeiffer, C. *Mental and Elemental Nutrients*, Keats Publishing Co., New Canaan, Connecticut, 1975
Roberts, S. E. *Exhaustion: Causes and Treatment*, Rodale Books Inc., Emmaus, Pennsylvania, 1967
Szekely, E. B. *The Essene Science of Life*, International Biogenic Society, Cartago, Costa Rica, 1978
Trowel, Hugh. *The Lancet*, 22 July 1978

Chapter 16
Schweigart, H. A. *Eiweis, Fette, Garzinfark*, Verlag H. H. Zanner, München

Recipe Section
Gerras, Charles. *Feasting on Raw Foods*, Rodale Press Inc., Emmaus, Pennsylvania, 1980
Hanssen, Maurice. *The Blender and Juicer Book*, Thorsons Publishers Ltd., Wellingborough, 1978
Kadans, Joseph M. *Encyclopedia of Medicinal Foods*, Thorsons Publishers Ltd., Wellingborough, 1979
McCallum, Cass. *The Real Food Guide Vol. 1 Fresh Fruit and Vegetables*, The Molendinar Press, Glasgow, 1981
de Nolfo, Joseph. *The Throw-out-your-stove-and-greasy-dishes No-cook Recipe Book*, Joseph's Rainbow, Oregon, 1976
Nuts and Seeds: the natural snacks, Rodale Press Inc., Emmaus, Pennsylvania, 1973
Petterson, Vicki. *Eat your way to Health*, Penguin Books, London, 1983
Reekie, Jennie. *Everything Raw*, Penguin Books, London, 1978
Schwartz, George. *Food Power*, McGraw-Hill, New York, 1979

Walker, N. W. *Diet and Salad Suggestions*, Norwalk Press, Arizona, 1940

Walker, N. W. *Raw Vegetable Juices*, Jove Publications Inc., New York, 1977

Whyte, Karen Cross. *The Original Diet: raw vegetarian guide and recipes*, Troubadour Press, San Francisco, 1977

Wigmore, Ann. *Recipes for Longer Life*, Rising Sun Publications, Massachusetts, 1978

INDEX

INDEX

Also available in Arrow by Leslie Kenton

THE BIOGENIC DIET

The Natural Way to Permanent Fat-Loss

SHED UNWANTED FAT WITHOUT SO MUCH AS
COUNTING A CALORIE!

How? By following the biogenic principles of living:

* Never mix concentrated proteins with concentrated
  starches

* Eat high-water raw foods like fresh fruit and vegetables,
  sprouted seeds and grains

* Heighten fat-burning through enjoyable exercise

The guidelines include suggested menus and over 100
exciting recipes. Switch over to biogenic living and feel
more alive than you ever have before!

Leslie Kenton, bestselling author, television broadcaster,
and Health and Beauty Editor of *Harpers & Queen*, is a
byword for health, beauty and fitness.

**'She is one of the best advertisements for her own medicine
you could hope to meet. When she strides into a room she
brings in the fresh air with her.' – *London Standard***

**'A shining and energetic example of the theory in practice.'
– *Observer***

AGELESS AGEING

The Natural Way to Stay Young

'She's a one woman Wall Street of well-being'
– *Cosmopolitan*

Famous for combining a high-tech approach to health and beauty with natural biological methods, award-winning writer and television broadcaster Leslie Kenton has devised a lifestyle that will prevent premature ageing, prolong vitality and good looks, and encourage longevity.

Readers are given the key to a personal anti-ageing plan that includes a complete body maintenance programme and advice on high-potency low-calorie nutrition and anti-ageing methods ranging from water therapy to the use of natural herbs and roots such as Siberian Ginseng. Looked at the Kenton way, staying young, healthy and good looking as the years pass is no longer an impossible goal.

ULTRAHEALTH

The Positive Way to Vitality and Good Looks

Want to feel good?

Look good?

Be at your peak both mentally and physically *all* the time?

Now you can, using a health programme that is practical, down-to-earth *and* really works.

Includes advice on:

* high energy nutrition
* self and environmental awareness
* age control
* physical fitness
* stress management

THE BOOK THAT WILL CHANGE YOUR LIFE

BESTSELLING HEALTH AND SELF-HELP TITLES

☐	No Change	Wendy Cooper	£3.99
☐	Understanding Osteoporosis	Wendy Cooper	£3.99
☐	The Vitamin and Mineral Encyclopedia	Dr Sheldon Saul Hendler	£8.99
☐	Feel the Fear and Do It Anyway	Susan Jeffers	£4.50
☐	Ageless Ageing	Leslie Kenton	£4.50
☐	The Joy of Beauty	Leslie Kenton	£8.99
☐	Ultrahealth	Leslie Kenton	£4.50
☐	Sexual Cystitis	Angela Kilmartin	£3.99
☐	Understanding Cystitis	Angela Kilmartin	£4.99
☐	Who's Afraid	Alice Neville	£5.99

ARROW BOOKS, BOOKSERVICE BY POST, PO BOX 29, DOUGLAS, ISLE OF MAN, BRITISH ISLES

NAME _____

ADDRESS _____

_____

_____

Please enclose a cheque or postal order made out to Arrow Books Ltd, for the amount due and allow for the following for postage and packing.

U.K. CUSTOMERS: Please allow 30p per book to a maximum of £3.00

B.F.P.O. & EIRE: Please allow 30p per book to a maximum of £3.00

OVERSEAS CUSTOMERS: Please allow 35p per book.

Whilst every effort is made to keep prices low it is sometimes necessary to increase cover prices at short notice. Arrow Books reserve the right to show new retail prices on covers which may differ from those previously advertised in the text or elsewhere.